英汉矿业翻译教程

主编◎王会娟　黄　敏　周建芝

A Textbook of

Mining Translation Between

English and Chinese

中国矿业大学出版社

· 徐州 ·

图书在版编目(CIP)数据

英汉矿业翻译教程 = A Textbook of Mining Translation Between English and Chinese / 王会娟，黄敏，周建芝主编. — 徐州 ：中国矿业大学出版社，2023.12

ISBN 978 - 7 - 5646 - 6130 - 4

Ⅰ. ①英… Ⅱ. ①王… ②黄… ③周… Ⅲ. ①矿业—英语—翻译—教材 Ⅳ. ①TD

中国国家版本馆 CIP 数据核字(2023)第 257535 号

书　　名　英汉矿业翻译教程
　　　　　Yinghan Kuangye Fanyi Jiaocheng
主　　编　王会娟　黄　敏　周建芝
责任编辑　万士才　吴学兵
责任校对　张梦瑶
出版发行　中国矿业大学出版社有限责任公司
　　　　　(江苏省徐州市解放南路　邮编 221008)
营销热线　(0516)83885370　83884103
出版服务　(0516)83995789　83884920
网　　址　http://www.cumtp.com　**E-mail**:cumtpvip@cumtp.com
印　　刷　江苏凤凰数码印务有限公司
开　　本　787 mm×1092 mm　1/16　**印张** 10.5　**字数** 203 千字
版次印次　2023 年 12 月第 1 版　2023 年 12 月第 1 次印刷
定　　价　38.00 元

前　言

当前,“新型全球化”要求我国行业和企业在全球化进程中发挥更大作用,也需要更多国际化人才的参与,“重塑全球化”。矿业作为我国国民经济重要的支柱产业之一,其技术发展水平、参与国际合作程度等都影响了整个行业的发展。因此,培养具备英语语言能力和沟通能力的矿业技术人才是当前国际化人才培养的重中之重。本教材即为普通高等学校矿业工程专业培养复合应用型人才而编写的,目标读者是英语专业本科生、翻译专业硕士、矿业工程专业学生以及有一定英语基础的矿业从业人员。教材从生产实践出发,选材广泛,将翻译理论、翻译技巧和矿业工程专业技术知识融合起来,兼顾了理论和实践的系统性、专业性和实用性。

本教材最大的特点是专业性强、适用面广。从矿业工程的专业视角给学习者提供鲜活的信息和语料,涵盖面广,语言素材选取涉及矿业工程的各个方面:矿井设计和开拓、井巷工程、岩体结构、矿井运输、通风、排水、事故防治、矿区环境保护和新型采矿方法等。读者通过本教材的学习可以储备相关矿业英语知识的语料,有助于拓宽专业知识面,建构复合型、宽厚型知识结构,为以后的矿业英语翻译、矿业行业国际合作交流打下良好的基础。

本教材另一个特点是实用性强,理论紧密联系实际。本教材概述了相关翻译理论,详细阐述了矿业文本实用的翻译方法和技巧,包括直译、意译、词类转换法、反译法、增词法、省略法、分译法、合译法、从句翻译以及矿业文本被动语态的翻译等,通过大量的矿业英语实例,深入浅出地介绍了矿业英语的词汇及句法特征,剖析运用不同的翻译技巧和方法,达到语言转换的目的。学生通过使用、阅读和学习本教材,能够接触到真实、鲜活的翻译实例,扩充自己的矿业工程知识;能够掌握相关翻译理论和技巧并通过实例操练,真正提高自己的矿业文本翻译水平。

本教材的整体结构合理,自成体系,操作性强。教材分为五章,第一章翻译概述,简要介绍翻译的性质、翻译的分类、翻译的标准以及矿业翻译的要求;第二章矿业文本的语言特点,介绍了矿业英语的词汇和句法特点;第三章矿业文本的翻译方法,包括直译和意译;第四章矿业文本的翻译技巧,通过具体译例,重点介绍了矿业文本的翻译技巧,包括增词法、省略法、词类转换法等;第五章矿业文本的翻译技巧综合练习,设置了矿业文本的单句翻译练习和篇章翻译练习。本教材讲解详细,便于教师讲授,易于学生参与,课堂互动性强,难易程度跨度大,适合各层次各水平的学生,便于教师组织课堂,操作性强。

在教材的编写过程中,我们得到了中国矿业大学外国语言文化学院宁淑梅老师、李冠杰老师、倪艳笑老师、吕淑文老师、潘荣成老师、房立政老师以及中国矿业大学矿业工程学院姬长生老师的大力帮助和支持;矿业工程学院马立强老师提供了很多翻译素材;此外,我们还参考借鉴了部分学者在科技翻译领域的研究成果,在此向他们表示衷心的敬意和感谢。

由于编写者经验和水平有限,教材中难免存在一些错误和不足之处,恳请广大读者批评指正!

编　者

目　录

第一章　翻译概述 …… 1

第一节　翻译的性质 …… 1
第二节　翻译的分类 …… 2
第三节　翻译的标准 …… 3
第四节　矿业翻译的要求 …… 5

第二章　矿业文本的语言特点 …… 8

第一节　词汇特点 …… 8
第二节　句法特点 …… 10

第三章　矿业文本的翻译方法 …… 14

第一节　直译 …… 14
第二节　意译 …… 17

第四章　矿业文本的翻译技巧 …… 22

第一节　增词法 …… 22
第二节　省略法 …… 30
第三节　词类转换法 …… 36
第四节　反译法 …… 45
第五节　分译法 …… 56
第六节　合译法 …… 69
第七节　从句的翻译 …… 75
第八节　被动语态的翻译 …… 93

第五章　综合练习…… 105

附录一：综合练习参考答案…… 120

附录二：矿业英语专业术语中英文对照…… 133

参考文献…… 156

第一章　翻译概述

第一节　翻译的性质

翻译是一门综合性的学科，因为它集语言学、文学、社会学、教育学、心理学、人类学、信息理论学等学科之特点于一身，在长期的社会实践中已经拥有了它自己的一套抽象的理论、原则和具体方法，形成了独立的体系，而且在相当一部分的语言材料中，这些方法正在逐渐模式化。视角的不同可以导致对翻译性质认识的差异，不同的翻译理论家对此有不同的定义。

翻译是用一种语言把另一种语言所表达的思维内容准确而完整地重新表达出来的语言活动。(张培基，1980)

翻译是一种跨越时空的语言活动，是"把一种语言已经表达出来的东西用另一种语言准确而完整地重新表达出来"。(范存忠，1985)

翻译的实质是语际的意义转换。(刘宓庆，1990)

翻译是将一种语言文字所蕴含的意思用另一种语言文字表达出来的文化活动。(王克非，1997)

Translating consists in reproducing in the receptor language the closest natural equivalent of the source language message, first in terms of meaning and secondly in terms of style. (Nida，1969)(翻译就是在译入语中再现与原语的信息最切近的自然对等物，首先是就意义而言，其次是就文体而言。)

Translation may be defined as follows: The replacement of textural material in one language (SL) by equivalent textual material in another language (TL). (Catford，1965) [翻译的定义也可以这样说：把一种语言(SL)中的篇章材料用另一种语言(TL)中的篇章材料来加以代替。]

Translation is a process in which the parole of one language is transferred into the parole of another with the content i. e. meaning unchanged. (Barhudarov, 1985)[翻译是把一种语言的言语产物在保持内容方面(也就是意义)不变的情况下改变为另外一种语言的言语产物的过程。]

A good translation is one which the merit of the original work is so completely transfused into another language as to be as distinctly apprehended and as strongly felt by a native of the country to which that language belongs as it is by those who speak the language of the original work.（好的翻译应该是把原作的长处完全地移注到另一种语言，以使译入语所属国家的本地人能明白地领悟、强烈地感受，如同使用原作语言的人所领悟、所感受的一样。）

Translation, being an intellectual activity, the object of which is the transfer of literary, scientific and technical texts from one language into another, imposes on those who practice it specific obligations inherent in its very nature.（翻译工作作为一种脑力劳动，其目的是将文学、科学方面的材料从一种语言译成另一种语言，从事这一工作的人承担着由该工作性质产生的特殊义务。）

第二节　翻译的分类

既然翻译的性质可从不同的角度来定义，那么同样翻译的种类也可从不同的视角来分类。一般说来，翻译可从五种不同的角度来分类。

(1) 从译出语和译入语的角度来分类，翻译可分为本族语译为外语及外语译为本族语。具体来讲，如英汉翻译、汉英翻译、德汉翻译、汉德翻译等。

(2) 从涉及的语言符号来分类，翻译分为语内翻译、语际翻译和符际翻译。语内翻译是指在同一种语言内部的翻译，如古代汉语译为现代汉语，即文言文本译为白话文本，唐诗译为现代诗，广东话译为普通话；又如古英语译为现代英语，英语诗歌译为英语散文等。符际翻译是指不同符号系统间的信息转换，通过非语言符号系统解释语言符号，或语言符号解释非语言符号，如交通信号和人类语言之间的转换（红灯—停止；绿灯—前行）；电报代码和人类语言之间的信息转换；把语言符号用图画、手势、数学、电影或音乐来表达等。语际翻译是指两种不同语言间的信息转换，把本族语译为外族语，或外族语译为本族语。译出的文本语言称为来源语，又称源语、译出语、出发语；译本呈现的语言为目的语，又称目标语或译入语。（此为翻译课的任务，或言际翻译。具体而言是英语、汉语之间的翻译。）

(3) 从翻译的手段来分类，翻译可分为口译、笔译和机器翻译。口译，顾名思义，是指译员以口语的方式，将译入语转换为译出语，分为同声传译和交替传译。笔译即笔头翻译，就是用文字翻译（区别于口译）。机器翻译又称为自动翻译，是利用计算机将一种自然语言（源语言）转换为另一种自然语言（目标语言）

的过程。它是计算语言学的一个分支，是人工智能的终极目标之一，具有重要的科学研究价值。

(4) 从翻译的处理方式来分类，翻译可分为全译、节译和编译。全译即完全翻译原文文本。节译指根据特定创作意图，仅从原作品中选择部分内容予以翻译，或摘选原文的主要内容进行翻译。编译就是将原文文本的大致意思用另一种语言表达出来，可以删减或增补。

(5) 从翻译的题材来分类，翻译可分为应用文体翻译、论述文体翻译、新闻文体翻译、文学文体翻译和科技文体翻译等。矿业文体翻译属于科技文体翻译的一种。

第三节　翻译的标准

翻译标准是指翻译实践时译者所遵循的原则。任何翻译实践总要遵循一定的翻译标准或原则，衡量一篇译文的好坏同样也离不开一定的翻译标准，因此翻译标准的确立对于指导翻译实践有着重要的意义。由于人们看待翻译的角度不同，自然有了不同的翻译标准。

我国清末启蒙思想家严复提出“信”“达”“雅”的翻译标准。他在《天演论》中的“译例言”讲到：“译事三难：信、达、雅。求其信，已大难矣，顾信矣不达，虽译犹不译也，则达尚焉。”“信”指意义不悖原文，即译文要准确，不偏离，不遗漏，也不要随意增减意思；“达”指不拘泥于原文形式，译文通顺明白；“雅”则指译文选用的词语要得体，追求文章本身的古雅，简明优雅。通俗地讲，译文要符合汉语言的特点，注重一句话的完整性，即内容、结构、文采，也就是内容的准确性，语法结构的顺畅，语言载体的文学性。

我国著名文学家、思想家鲁迅在《且介亭杂文二集・题未定草(二)》一文中指出：“凡是翻译，必须兼顾着两面，一当然力求其甚解，一则保存着原作的丰姿。”他的这一主张，后来得到了人们的普遍认可，因为它与我们现行的翻译标准——“忠实”和“通顺”，事实上是同一含义，至今仍有重要的现实意义。鲁迅还指出：凡是翻译，“动笔之前，就先得解决一个问题：竭力使它归化，还是尽量保存洋气”。鲁迅这里所说的“保存洋气”，其实就是“信”，是使译作保存“异国情调”，“保存着原作的丰姿”。

18 世纪末的英国学者亚历山大・弗雷泽・泰特勒(Alexander Fraser Tytler，1747—1814)在《论翻译的原则》(*Essay on the Principles of Translation*)一书中提出了著名的翻译三原则。

(1) 译文应完全复写出原作的思想。(A translation should give a complete

transcript of the ideas of the original work.）

(2) 译文的风格和笔调应与原文的性质相同。(The style and manner of writing should be of the same character as that of the original.）

(3) 译文应和原作同样流畅。(A translation should have all the ease of the original composition.）

傅雷以翻译法国文学作品享誉译坛，其在《高老头·重译本序》中提出“神似”说：“重神似而不重形似；得其精而忘其粗，在其内而忘其外”“以效果而论，翻译应当像临画一样，所求的不在形似而在神似。以实际工作而论，翻译比临画更难”。

1964年钱钟书在《林纾的翻译》一文中提出“化境”说。他认为文学翻译的最高理想可以说是“化”。把作品从一国文字转变成另一国文字，既能不因语言习惯的差异而露出生硬牵强的痕迹，又能完全保存原作的风味，那就算得入于“化境”了。

林语堂是当代著名学者、文学家、语言学家。他提出翻译的标准问题大概包括三方面：第一是忠实标准，第二是通顺标准，第三是美的标准。

许渊冲提倡文学翻译要做到“意美、音美、形美”(beauty in meaning, beauty in sound and beauty in form)，并在这一理论的指导下译出了举世公认的优秀文学译作。翻译原则的共同特点可以说是译文重神似非形似，语言必须美，即许渊冲先生所主张的“words in best order”“best words in best order”[英国19世纪诗人塞缪尔·泰勒·柯尔律治(Samuel Taylor Coleridge)]。

现在我们把翻译标准笼统地概括为“忠实、通顺”四个字。所谓忠实，首先指忠实于原作的内容。译者必须把原作的内容完整而准确地表达出来，不得有任何篡改、歪曲、遗漏或任意增删的现象。忠实还指保持原作的风格——原作的民族风格、时代风格、语体风格、作者个人的语言风格等。译者对原作的风格不能任意破坏和改变，不能以译者的风格代替原作的风格。总之，原作怎样，译文也应怎样，尽可能保留本来面目。也即鲁迅说的，翻译必须“保存着原作的风姿”。所谓通顺，即指译文语言必须通顺易懂，符合规范。译文必须是明白晓畅的现代语言，没有逐词死译、硬译的现象，没有语言晦涩、文理不通、结构混乱、逻辑不清的现象。鲁迅所说的，翻译必须“力求其易解”，也就是这个意思。例如：

1) Since high voltage will exist in many centrifugal pump rooms it will be wise to place a gate across the entrances, and this should be kept locked. Warning signs should be placed on the gate. Screens at the sides and backs of switchboards are available.

由于许多离心泵的泵房都使用高压电，所以最好在进出口处设门并加锁。

门上应设警报信号。在配电盘的两侧和背面最好加设屏板。

2）加强煤场管理，禁止乱堆乱放。

Strengthen the management of coal yard and prohibit clutter.

以上两例在最大限度地忠实再现原文的信息和内容的同时，对原文本进行了不同程度和方式的调整，尽量做到通顺、地道、达意，使译文读者更容易理解译文。

忠实与通顺是相辅相成的。忠实而不通顺，读者看不懂，也就谈不上忠实；通顺而不忠实，脱离原作的内容与风格，通顺也失去了作用，使译文成为编纂、杜撰或乱译。

目前，就科技翻译包括矿业文本的翻译而言，还没有一个公认的统一标准。因此根据翻译的本质和翻译理论以及语言的特点，在总结前人提出的翻译标准的基础上，总结出下列矿业文本的翻译标准。

（1）准确，即理解和表达科技内容，包括科技概念（尤其是科技术语）、语言形式、逻辑关系、符号公式、图表数字等要准确无误，要忠实于原文。

（2）简洁，即用词、造句、行文要简洁明了，精练通顺。

（3）规范，即语言、文字、术语、简称、符号、公式、语体、文章体例、计量单位等都要规范统一，符合国家标准和国际标准。

第四节　矿业翻译的要求

要做好矿业文本翻译工作就要具备良好的业务素质。良好的业务素质指的是扎实的语言功底、出色的写作技能、一丝不苟的工作作风、丰富的专业知识以及过硬的翻译理论知识和熟练应用翻译技巧的能力。具体说来，这些业务素质至少包括四个方面的内容。

第一，充分了解矿业文本的语言特点。矿业文本无论是在词汇层面还是在句法层面都有其特点，译者应充分了解这些特点，才能更好地理解原文，进而更准确完整地进行翻译。例如：

1）When the top cage is loaded, the man called the bankman signals to the engineman at the winding house.

罐笼安装好之后，煤炭采掘工发送信号给绞车房司机。

上例中包含有关采矿工程相关专业词汇，如 top cage 应为“罐笼”，bankman 应为“采掘工”，the winding house 为“绞车房”，这些词汇均应准确理解和翻译。

2）Shield supports are designed to provide great resistance to lateral (horizontal) movement of roof strata, due to the method of linking the shield to the base; this saves the legs from damage.

由于采用盾构与底座连接的方法,盾构支架的设计可为顶板岩层的横向(水平)移动提供巨大阻力,这样可以避免主柱受损。

原文中的 shield supports are designed to provide great resistance 为被动语态结构,而被动语态是科技文本,特别是矿业文本中句法层面的主要特点。翻译时应进行具体分析,恰当转换成符合目的语规范的译文。

第二,具有丰富的矿业工程相关知识。这有助于正确理解原文,准确地表达原作所要传达的内容。例如:

1) According to the depth and diameter of different borehole, drilling and blasting method can be divided into shallow hole blasting method, medium-depth hole blasting method and deep-hole blasting method.

根据炮眼深度与直径的不同,钻眼爆破法分为浅孔爆破法、中深孔爆破法和深孔爆破法。

2) The quality of acid mine water is complex. At present the main processing methods are neutralizational process, biochemistry, reverse osmosis, etc.

酸性矿井水水质复杂,目前主要的处理方法有中和法、生物化学法、湿地生态工程法、反渗透法等。

例 1)是有关钻眼爆破法的处理方式,例 2)涉及酸性矿井水水质处理的相关方法,如果译者对相关矿业知识不够了解,就很难理解原文,更谈不上准确地道地再现原文了。

第三,熟悉翻译理论和常用翻译技巧和方法,善于灵活运用各种翻译技巧。要提高翻译质量,译者还必须在实践中加强翻译理论和技巧的学习和研究,不断练习总结,逐步掌握各种翻译技巧和方法,提高翻译能力。例如:

1) The long-distance separate operation of human and machine has come true.

实现人、机分离远距离操作。

此句的翻译中,采用了逆译法和语序调整法。

2) The majority of installations are electrically driven. Alternating current is used most often, with the voltage ranging from 200 up to 400 volts.

大多数设备由电力驱动。交流电最常用,电压 200～400 伏。

翻译本例时,译者首先采用词类转换法,把原文的名词 majority 转换成汉语的形容词“大多数”,来修饰“设备”。翻译时,译者也用到了省略法,省略了 ranging from 等词。

第四，具有严谨认真的态度。翻译过程中，译者认真不认真、严谨不严谨，译文的质量差别是比较大的。翻译的每一个步骤都必须严肃认真，否则理解不准确、表达时粗心大意、译后也不认真检查和审校，译文质量可想而知。事实上，在动手翻译之前，要仔细揣摩矿业文本的主要内容和文体特点，准确完整再现原文的信息。翻译完成之后要认真仔细检查，发现问题，及时认真修改和润色，这样才能有一个忠实和通顺的译文。

第二章 矿业文本的语言特点

科技英语是指在自然科学和工程技术领域使用的英语，一般包括科技论著、科研论文、研究报告、技术标准等，其主要作用是表达和传递科技知识和信息。科技英语在句法结构和词汇方面都形成了其特有的习惯用法。它要求以最少的文字符号准确地表达、传递最大的信息量，因此科技英语的特点可以概括为准确性、客观性和简明性。矿业工程属于工程技术领域，包括矿山勘探、采矿作业、设备操作规程、矿井建设、矿山安全等方面。作为科技英语的重要构成部分，矿业英语在词汇和句法等方面具有鲜明的特点。

第一节 词 汇 特 点

在矿山开采和矿产加工生产等过程中，相关工程术语必须准确、客观和专业。矿业英语的词汇特点主要表现在三个方面：大量的专业词汇、借用普通词汇以及灵活的构词方式。

1. 大量的专业词汇

矿业英语的专业词汇主要是关于选矿、采矿、物质分析、加工等一系列工业操作流程的词汇，是指只有一种专业含义的词汇，包括在实践中根据需要创造出来的词汇，如表 2.1 所列。

表 2.1 矿业英语专业词汇举例

专业词汇	专业词义	专业词汇	专业词义
anthracite	无烟煤	underseam	下部煤层
burgy	煤屑	bugdusting	清除煤粉
coom	煤烟	coyoting	不规则的小窑开采
slag	矿渣	fandrift	扇风机引风道
tailings	尾矿	gatehead	装载点
backlye	井下错车道	dressing machine	选矿机

表2.1(续)

专业词汇	专业词义	专业词汇	专业词义
pelite	泥质岩	durain	暗煤

2. 借用普通词汇

矿业英语中借用了很多普通词汇，并赋予其新的专业含义。因此，在理解和翻译这类词汇时需严谨，应结合矿业的专业背景知识和具体语境，必要时需借助专业词典，从而准确把握其语义。如表 2.2 所列。

表 2.2　普通词汇表达专业含义举例

词汇	专业词义	普通词义
cope	上(砂)箱	处理，对付
roadway	巷道	道路，路面
booze	铅矿	豪饮，痛饮
blank	坯料，板	空白的，空虚的
throat	排矿口	咽喉
bootleg	哑炮，拒爆炮眼	非法制造并销售

3. 灵活的构词方式

矿业科技的发展带来了相关矿业英语的发展，表现在词汇层面，就是其灵活多变的构词方式。

(1) 合成法。一般将两个或两个以上的词组合在一起，构成一个新词，这种构词法叫合成法，也是矿业英语术语扩展的一个重要方法。在矿业英语中，合成法主要构成的是复合名词，主要构词形式有：名词＋名词、介词＋名词、形容词＋名词等。例如：

airway 风巷　　　　　blasthole 爆破孔

airwinch 风动绞车　　catchpit 排水井

intake 进风巷道　　　longhole 深炮眼

(2) 混成法。将两个词混合或各取一部分缩略成一个新词，前半部分表属性，后半部分表主体，这样的构词法就是混成法。因为各个学科间的交叉发展，矿业英语里也出现很多的混成词。例如：

greentech＝green technology 绿色科技

foamcrete＝foam concrete 泡沫混凝土

tectonogram＝tectonics diagram 图解

metallogenic＝metallogenesis genetic 成矿的

gravisphere＝gravity sphere 引力作用范围

(3) 词缀法。词缀法是矿业英语中非常重要的一种构词方法，也叫派生法，即在词根上加前缀或者后缀，构成新的单词的方法。很多英语词缀大都来源于希腊语和拉丁语，有很强的构词能力，掌握好词缀法，能够大大地扩充英语词汇量。常见的前缀有：anti-(反对，相反)，ir-(不，无)，counter-(相反)，de-(去掉，离开)；常见的后缀有：-meter(仪，表)，-ology(……学)等。例如：

dehydrogenation 去氢　　antifreeze 防冻剂

counterlode 交错矿脉　　irreversibility 不可逆性

anthracology 煤炭学　　viscometer 黏度计

(4) 缩略词。矿业英语中缩略词的使用也比较常见，因为缩略词使用起来简便快捷，意义表达精确，缩略词的构成方法最常见的是首字母缩略。例如：

RQD(rock quality designation index)岩石质量指数

HDR(hot-dry rock)干热岩

ASN(average sample number)平均抽样数

TBM(tunnel boring machine)隧道掘进机

FTA(fault tree analysis)事故树分析

第二节　句法特点

矿业英语主要是描述矿山开采、生产的过程以及相关的科学研究结果，要求客观、准确地阐述描述对象的特性、规律、研究方法和研究成果等。矿业英语文体严谨、句式严整、行文简练、逻辑性强，其句法特点主要表现在以下四个方面。

1. 大量使用名词化结构

名词化结构是指把句中的动词或形容词等转换成名词或名词词组的结构，它能高度概括事物的本质，简练表达复杂的思想，因而在矿业英语中，名词化结构使用的频率很高，其中由动词派生出来的名词化结构最为常见，因为它们既保留着动词的特点，又兼有名词的功能，搭配力比较强。例如：

1) Combinations of controlled blasting techniques are used.

各种控制爆破可以组合起来使用。

2) Core examinations can be supplemented with borehole observations made with either an optical or television borehole camera.

岩芯的测定工作可以辅以下列手段——利用光学或电视摄像头对钻孔进行观察。

3) The reclamation laws require revegetation of mined areas and restoration of disturbed areas to an acceptable post-mining land use.

复垦法规要求采后矿区应种植植被和复垦,并应使采后土地利用达到可接受的水平。

2. 常使用现在时态,包括一般现在时和现在完成时

矿业英语中经常涉及很多的科学概念、定理、定义、公式和图表,以及要对客观的生产、实验过程进行精确描述,这些内容一般不因时间因素的影响而变化,所以采用一般现在时更能客观、准确地表达信息。现在完成时也属于现在时的范畴,常常用来总结已经取得的研究成果和描述已经发生的现象。例如:

1) Supported mining methods are often used in mines with weak rock structure.

充填采矿法一般在岩体不稳固的矿山使用。

2) Significant revisions have required in these processes to ensure that the environment is protected and pollutional discharges minimized, while maximizing mineral recovery.

需要对这些流程作重大改进,以确保环境得到保护,污染物排放最小化,同时矿物回收率最大化。

3) Coal exploiting has caused a lot of ecological and environmental problems including land and vegetation destruction, water and air pollution, heavy metal pollution and so on.

煤炭开发已经造成了很多生态和环境问题,包括土地和植被破坏、水和空气污染、重金属污染等。

3. 常使用被动语态

矿业英语所描述的主体通常是自然现象或客观事物,并且经常涉及计算公式,因此采用被动语态更能强调所述事物本身,比主动结构主观色彩更少,使行文更加简洁流畅。例如:

1) Wall and back support may be accomplished by leaving random or even systematic pillars.

可以通过留不规则矿柱甚至留规则矿柱来支护围岩和顶板。

2) The energy generated by the booster fan is used to overcome the resistance of the split in which it is installed.

局部通风机产生的能量被用来克服它所在分风线路上的阻力。

3）The unsupported methods of mining are used to extract mineral deposits that are roughly tabular and are generally associated with strong ore and surrounding rock.

空场法用于开采板状矿体，通常要求矿体和围岩比较稳固。

4. 常使用复杂的长句结构

为了客观准确地表达复杂的概念，突出鲜明的逻辑关系，矿业英语中常常使用包含很多修饰语、并列结构和从句的复杂长句。要正确理解其逻辑含义，必须理清各个语法成分之间的关系。例如：

1）Furthermore，even where mineralization extends to a greater depth in open pit mines，the rapidly increasing amount of overburden to be handled imposes economic limits beyond which mining must either be abandoned or converted from surface to underground operations.

另外，如果露天开采的矿体向下延深很多，下部急剧增加的待处理废石剥离量会引出经济合理开采极限问题，超过了这个极限，要么结束开采，要么就要由露天开采转为地下开采。

本句是个复杂长句，语法成分复杂：where 引导的地点状语从句表示地点；不定式短语 to be handled 作 overburden 的后置定语；介词 beyond 后面是 which 引导的定语从句修饰前面的 limits。为了准确理解句子意思，必须理清各成分之间的逻辑关系。

2）Even where the development of a mineral reserve is not prohibited by environmental legislation，added environmental costs imposed on mineral production can have a direct and significant impact on the domestic industry's viability，particularly where foreign competitors are not subject to the same or similar economic burdens.

甚至在矿产开发不受环境法制约的地方，对矿产开发征收的高额环保费，特别是在国外同行竞争者并没有承受同样或类似的经济负担的情况下，会对国内产业的生产能力产生直接而严重的冲击。

本句是个复合句，第一个 where 引导的地点状语从句表示“在……地方”，过去分词短语 imposed on mineral production 作后置定语修饰前面的 costs，表示“被施加在……上”，第二个 where 引导的也是地点状语从句表示地点。

3）In underground mining，the walls and the roof or back of openings are not usually self-supporting，although where ore bodies have a quartzite，strong limestone or other competent roof，some very large openings have been excavated open without the aid of artificial support.

在地下开采中，巷道的两壁和顶板通常要进行人工支护，虽然有些地方的矿体是石英岩、坚固的石灰岩或其他顶板，但有些大的断面巷道是在没有人工支护的情况下开凿出来的。

本句是一个复杂长句，句中有一个 although 引导的让步状语从句，一个 where 引导的地点状语从句说明地点，还有 without 引导的介词短语作状语。

第三章　矿业文本的翻译方法

在矿业英语的翻译过程中，最常用的翻译方法为直译（literal translation）和意译（free/liberal translation）。直译与意译是最基本也是最早为人们所关注的翻译方法，是非常有代表性和概括性的两种翻译方法。国内外翻译界自古至今一直都有直译、意译之争，美国翻译家奈达以及我国的翻译家傅雷提倡意译；英国的纽马克和我国的鲁迅、瞿秋白则主张直译。从翻译实践来看，直译与意译是相互关联、互为补充的，同时，它们又互相协调、互相渗透、不可分割。

第一节　直　　译

所谓直译就是既保留原文内容、又保留原文形式的翻译方法。具体来讲，直译以句子为基本单位，在翻译过程中要兼顾整个文本。直译力求再现原作的思想内容和写作风格，尽量保留原文的修辞手法和主要句型结构。译文(或目的语)常常使用相同的表达形式来体现原文(或源语)的内容，并能产生同样的效果。在这种情况下，我们就采用直译。例如：

1) Most mineral deposits have been geometrically characterized as to an idealized shape, inclination, size, and depth. Complex or composite bodies are then composed of these elements.

大多数矿物矿床的几何形状都具有理想的形状、倾角、大小和深度。复杂或复合的主体由这些元素组成。

此例的翻译基本保留了原文的句型结构，大多词性也没有改变，只是在翻译第二个短句时，在后半部分作了语序的微调，即将 then 提前至该句句首，composed of these elements 语序调整为“由这些元素组成”。译文既忠实亦通顺，是直译的典型。

2) European governments require that all possible deposits be mined to conserve the nation's energy resources.

欧洲各国政府要求采出所有可采的煤层以保护国家能源。

此例译文完全保留原文的句型结构，只是中间部分 all possible deposits be

mined 在翻译成汉语时微调语序为"采出所有可采的煤层",原文内容也全部再现。

3) The walls will eventually be lined with concrete so that when the shaft is thousands of feet deep it will be supported by a continuous concrete tube.

井壁最后用混凝土衬砌,以便竖井凿到数千英尺深时,有一整体的混凝土管支撑。

原文是英语主从复合句,包含一个主句、两个从句。两个从句分别为 so that 引导的目的状语从句以及 when 引导的时间状语从句。翻译时,基本按照原文的语序进行,原文内容完全保留,译文也比较通顺达意。

4) Periodic reworking and repair is necessary.

定期返修是必要的。

该例为矿业翻译中典型的直译案例。译文完全保留了原文的句型结构、语序和词性,内容忠于原文,译文地道通顺。

5) With the introduction of machines to expedite and cheapen the cost of headings, these disadvantages are not so serious as in the past.

由于采用机器开掘煤巷,降低了成本,缩短了工期,这些劣势已不像过去那么严重了。

6) The plow head is equivalent to the cutting drum in the shearer, but it contains no motorized equipment.

刨头相当于采煤机的切割滚筒,但是它不包含电动设备。

7) Generations of this vegetation died and settled to the swamp bottom land over time the organic material lost oxygen and hydrogen, leaving the material with a high percentage of carbon.

这些植物一代又一代枯死后沉积到沼泽底部,在漫长的岁月中,这些有机(植)物失去了氧元素和氢元素,留下了高含量的碳元素。

8) The main reasons include: ① air exists in the pump; ② the fluid volume in the fluid tank is insufficient; ③ the pipette is not well sealed; ④ the nitrogen pressure of the accumulator is insufficient; ⑤ the one-way valve core of the suction and discharging valve is not well sealed or the lifting height is of inconformity.

主要原因包括:① 泵体内存有空气;② 液罐内液量不足;③ 吸液管密封不严;④ 蓄能器充氮压力不足;⑤ 吸排液单向阀阀芯密封不良或提升高度不符合要求。

9) Such fans were noisy at high pressures and were not particularly

efficient, but their simplicity, elasticity in operation and cheapness led to investigations which culminated in the modern all-metal fan with guide vanes streamlined casings and blades of aerofoil section capable of adjustment for variable pitch to meet the requirements of the mining industry and giving high efficiencies.

这种通风机在高压下噪声很大，效率不是很高，但是它结构简单、使用灵活、价格低廉，引起人们进一步钻研的兴趣，最终产生了有导流片的现代全金属结构通风机，这种通风机具有流线型风道，翼型截面的叶片，它可以调整螺距，以满足采矿工业的各种需要，效率很高。

10) Coal formation continued throughout the Permian, Triassic, Jurassic, Cretaceous and Tertiary Periods, which spanned 290 million to 1. 6 million years ago.

煤的形成还依次跨越了二叠纪、三叠纪、侏罗纪、白垩纪和第三纪，大约从二亿九千万年前到一百六十万年前。

以上例子基本采用直译的方法，既保留原文的意义又最大限度保留原文的结构，是比较典型的直译范例。

11) 平顶山煤业(集团)有限公司(以下简称平煤集团)位于河南省平顶山市。其前身是平顶山矿务局。

Pingdingshan Coal (Group) Co. Ltd. [hereinafter referred to as Pingdingshan Coal (Group)] is located in Pingdingshan City, Henan Province. Its predecessor was Pingdingshan Mining Administration Bureau.

该例翻译完全保留汉语的句式结构和语序。英语译文既忠实又通顺。

12) 三轴强度试验通常在三轴压力室内对圆柱形试件进行加载。围压由液压提供，而纵向载荷由试验机施加。

The triaxial strength test is usually performed for a cylindrical specimen in a triaxial cell. The confining pressure is applied by using hydraulic oil, whereas the vertical load is applied by the testing machine.

该例原文由两个句子构成，第二句为转折复句，由“而”引导。翻译成英文时完全保留了原文的句型结构，转换成英语的两个句子，第二句为英语主从复合句，个别语序作了微调。

13) 中国矿产资源承载能力研究较早，但由于侧重点不同，对于受环境影响的煤炭资源承载力研究相对较弱。

The research on the bearing capacity of mineral resources in China is relatively early, but due to the different emphasis, the research on the bearing capacity of coal resources affected by the environment is comparatively weak.

14）为分析、评价我国矿业类高校本科生科研创新能力，以具有代表性的四所矿业高校为例，本论文综合分析影响科研创新能力的相关因素。

In order to analyze and evaluate the scientific research and innovation ability of undergraduate students in mining universities in China, taking four representative mining universities as examples, this paper comprehensively analyzes the relevant factors affecting the scientific research and innovation ability.

15）1996年1月，经煤炭部批准，平顶山矿务局重组为国有独资有限公司。2002年12月，通过债权置换，重组为股份有限责任公司。

In January 1996, by approval of the former Ministry of Coal Industry, Pingdingshan Mining Administration Bureau was restructured into the state-owned and sole proprietor Co. Ltd. and in December 2002, by debt-to-equity swap, it was restructured as the limited liability company with multiple stock owned.

以上例句的翻译采用了直译方法，较好地保留了原句的句型结构，标点符号也基本与原文一致，原文的内容充分再现，只是个别地方的语序进行了微调，译文既忠于原文，又通顺达意。

第二节　意　　译

所谓意译就是只保持原文内容、不保持原文形式的翻译方法。意译可定义为一种补充手段。译文的语言(或目的语)与原文的语言在许多情况下没有同样的表达形式来体现同样的内容，也不能产生同样的效果。在这种情况下，一般采用意译较好。例如：

1）The extraction and processing of minerals is an essential part of the way the world and its various civilizations function and interact.

矿物的开采和加工是世界及其各种文明发挥作用和相互作用的重要组成部分。

在上句的翻译过程中，涉及英汉两种语言关于定语位置的差异。英语定语从句均为后置，而汉语修饰语大多为前置，因此，上句中 the world and its various civilizations function and interact 在译成汉语时调整至“重要组成部分”之前，这样更加符合汉语表达习惯，目的语读者更容易接受。

2）Indeed, such is the nature of the ultra-deep mines in South Africa that even today this country's mining engineers are considered the real pioneers when it comes to solving the problems associated with deep-level mining.

事实上，南非超深井的性质就是如此，即使在今天，该国的采矿工程师在解决与深度开采相关的问题时也被认为是真正的先驱。

在上例的翻译中，对原文的句型结构作了调整，把从句部分 when it comes to solving the problems associated with deep-level mining 向前提，使译文更易被目的语读者理解和接受。

3) The majority of mining companies have made massive advances to address both environmental and social issues. Nevertheless, the extraction of natural resources now attracts the sort of responsibility and scrutiny that few other global industries are subjected to.

大多数矿业公司在解决环境问题和社会问题方面取得了巨大进展。然而，如今开采自然资源需要承受的责任和面临的审查比全球其他行业更大、更严得多。

上句的翻译在忠实再现原文内容的基础上，对语序进行了调整，使其更加符合汉语的习惯。

4) The geophysical techniques that can be applied to the investigation of coal deposits and the improvement in controlling the coal mines surrounding rock include seismic method, resistivity method and electromagnetic method.

地球物理技术包括地震法、电阻率法和电磁法，可用于煤矿床调查和改进煤矿围岩控制。

此句的翻译涉及 that 引导的定语从句的译法。关于英语定语从句的翻译，一般有两种翻译方法：其一，翻译成汉语前置修饰语；其二，采用分译法，将英语的后置定语分译成独立的一个部分或单句。此句中，将 that can be applied to the investigation of coal deposits and the improvement in controlling the coal mines surrounding rock 分译成一个独立的部分“可用于煤矿床调查和改进煤矿围岩控制”，这样译文既清晰明了又通顺达意。

5) For these reasons some kinds of support for the surrounding rock is often required in underground excavations which are to be kept open to enable mining operations to continue.

由于这些原因，地下挖掘通常需要对围岩进行支撑，支撑的地方应保持畅通，以使采矿作业能够顺利进行。

6) Coal mining technology refers to the methods, equipment and their coordination in time and space used in all process at the coal face.

采煤工艺是指采煤工作面各工序所用的方法、设备及其在时间、空间上的相互配合。

在上面两个句子的翻译中，均对原文的句型结构进行了调整，具体采用了分译和语序调整等翻译技巧，使译文地道通顺。

7) Ramp Plates are fitted to the face side of the line pans and designed to scrape the floor clean as the structure is advanced.

铲板装于机槽的工作面一侧，其设计目的是在推移输送机时将底板铲刮干净。

上例对原文中的时间状语 as the structure is advanced 的语序进行调整，使译文不仅忠实于原文的内容，而且通顺易懂。

8) It is very important to prevent coal dust accumulating because it is a serious fire hazard.

因为煤尘是严重的火灾根源，因此防止煤尘积聚非常重要。

9) It is very significant to learn some experience and strategies through studying the development process of shale oil upstream industry in USA.

研究美国页岩油上游产业发展进程，学习其发展策略和经验，对于探索我国页岩油等非常规油气产业发展路径十分重要。

英汉两种语言存在巨大差异，在句法结构方面也是如此。英语一般先总结后叙述，而汉语正好相反，一般在句首先解释或阐释而后总结。因此翻译这类句子时，要注意句型结构的调整。以上两个例子均涉及此方面的调整。以上的调整，把英语中 it is very important 和 it is very significant 调整至汉语译文的尾部，把解释和说明部分放在句子的前面。

10) 绿色矿山建设是矿业绿色发展的路径选择和必然要求，是适应我国特殊资源国情和矿业特定发展阶段的现实选择。

Green mine construction is a path choice and inevitable requirement for the green development of the mining industry. It is also a realistic choice to adapt to China's special resource situation and the specific development stage of the mining industry.

上例是有关中国绿色矿山建设的内容。翻译时对汉语原文进行了结构调整。首先使用分译法，在“必然要求”之后断句，把原文分译成两个句子；然后把原文中结尾部分的“现实选择”放在第二句的句首，增加形式主语 it。由于英汉两种语言的差异，在翻译此种句型时我们只能舍弃外形而保留内容了。

11) 中国对铁矿石的进口总量一直增长，已成为全世界商业铁矿石、铜、铝、镍、钢和煤最大的消耗国，从 2010 年到 2016 年，中国的铁矿石进口量有望翻一番。

China's iron ore imports are expected to double from 2010 to 2016, following many years of growth that has made China the world's largest consumer of traded iron ore, copper, and aluminum, together with nickel, steel, and coal.

英语为形合的语言，而汉语为意合的语言。汉语以小短句见长，英语中长句和复合句居多。鉴于此，上句的翻译对原文的结构进行了调整，把原文中的四个小短句合译为英语的一个长句。

12）文章在研究矿业发达国家矿山环境治理模式基础上，结合我国国情研究提出中国建立矿山生命周期性环境治理体系建议。

Based on the study of mine environmental governance models of the developed mining countries, according to the study of our national conditions, this article puts forward suggestions on the establishment of an environmental governance system of mine life cycle in China.

上例翻译中，把原文中“在研究矿业发达国家矿山环境治理模式基础上，结合我国国情研究”部分放在句首，分别翻译成分词短语和介词短语，在译文中充当方式状语，然后再翻译主句。

13）壁式采煤法工作面长，一般100～200米，可以容纳功率大、生产能力高的采煤机械，因而产量大、效率高。

The wall type coal mining method has a very long working face of about 100 to 200 meters. The working face can hold the coal mining machinery of large power and production capacity and therefore high output and efficiency are achieved.

此句对原文的结构进行了整合，主要使用了分译法。

14）在矿井中，可以通过各种方法把煤从煤层中开采出来，比如常规采煤法、连续采煤法、长壁开采法和房柱开采法。各种开采形式都需要支护。

In the mine, the coal is extracted from the seam by various methods, including conventional mining, continuous mining, longwall mining, and room-and-pillar mining. Support is required for any opening.

15）采煤工作面昼夜循环次数和循环进度总称为循环方式，其种类有三种：单循环、双循环和多循环。

The number of day and night cycles and cycle progress of mining face are called cycle mode. Three categories are available, that is, single cycle, double cycles, and multi-cycles.

此例翻译中调整“昼夜循环”的语序，增加 that is。

在矿业文本翻译实践中，译者应具体问题具体分析，灵活地采用直译、意译和直译意译结合的方法进行翻译，直译与意译应该互相兼容和补充。直译与意译争论的核心在于如何处理形式、内容和目的语读者接受程度的辩证关系。对原文忠实是直译和意译的共同出发点和归宿。直译强调通过再现原文形式，实现在形式和内容两个层面对原文的忠实。目的语读者的接受程度也在直译考虑之列，但并非决定性因素。意译将目的语读者的接受程度作为衡量译文是否忠实于原文内容的重要标准，原文的形式则不在考虑之列。即在必要时，可以完全舍弃原文形式，而致力于用目的语读者完全能够接受的译文，忠实传达原文内容。

第四章　矿业文本的翻译技巧

第一节　增 词 法

由于英汉两种语言之间存在巨大差异，在翻译中为了准确地传达出原文的信息，译者往往需要对译文作一些增删。增词法就是在翻译时按意义上（或修辞上）和句法上的需要增加一些词来更忠实通顺地表达原文的思想内容。但是这种增词不是对原文内容的随意增词，而是增加原文中虽无其词却有其意的一些词。

一、增补原文中省略的词语

1) Where sinking stages are employed, the stage is raised prior to blasting to sufficient height from the shaft bottom to the removable mechanical cleaning unit and other stage equipment beyond damage distance.

在采用凿井专用吊盘的地方，爆破前，靠近底部的吊盘要提升到从井筒底部到可移动的机械清理设备之间的足够高度，其他吊盘设备也要提升到危险距离之外。

译文中增补了原文因上下文省略的“也要提升到”。

2) The main tunnels are lit by electric lights and so is the coal-face.

井下主巷道用电灯照明，采煤工作面也采用电灯照明。

原文中，so is the coal-face 为省略句，译文中增补了原文省略的“采用电灯照明”。

3) Single or double drums similar in some respect to those used in continuous miners but larger in diameter are mounted on the face side of the shearer loader.

安装在滚筒采煤机靠近工作面一侧的单滚筒或双滚筒，在某些方面与连续采煤机上的单滚筒或双滚筒相似，但直径大一些。

4) 将 31 煤采高和 32 煤综合采高代入式 2，得到 31509 工作面和 32509 工作面开采后覆岩导水裂隙带最大高度分别为60.7米和88.6米。

Separately substituting the mining height in seam #31 and the integrated mining height in seam #32 into Eq. 2 indicates that the maximum heights of the HCF zone at working faces 31509 and 32509 are 60.7 m and 88.6 m, respectively.

5）随着开采空间的增大，工作面后方采空区中部的部分覆岩裂隙逐渐被压实闭合，降低为56米。

As the goaf expanded, fractures over the middle of the goaf partially closed due to compression, and the fracture height was reduced to 56 m.

译文中增补了原文因上下文省略的“裂隙高度”。

二、增加语义、语法和修辞方面需要的词语

（一）增添必要的连接词

1）The chain unit must be very strong, rigid, and highly wear resistant because it is subjected to heavy external loading, both dynamic and static, and needs to overcome frictional resistance.

链条装置必须结实坚固，具有刚性，并有高度的耐磨损性能，因为无论在运动状态，还是在静止状态，它都要经受沉重的外载，而且需要克服摩擦阻力。

因逻辑需要，译文中增补了“无论……还是……”“都”。

2）If the conveyor needs to be shortened, the moving car is pulled by the winch toward the tail end, causing the run belt at the tail end to move toward the head end and storing excess belt in the belt bank.

如需缩短带式输送机，则用小绞车将可移动的张力车拉向机尾端，从而使在机尾端运行的输送带向机头端移动，并将多余的输送带储存进储带段。

3）Flow caused by unequal densities or weights of air columns in or near the openings (due mainly to temperature differences) is “natural draft” flow, and resulting pressure differences are “natural draft” pressure.

在井口内或井口附近，由于空气柱的不同密度或重量所引起的风流（主要是由于温差所致），称之为“自然风流”，所造成的压差，称之为“自然风压”。

4）By counting the lamps that are given out each day, the lamp-man knows how many miners are down the mine.

通过计算每天发放的矿灯数，就能知道有多少矿工下井。

5）With wire-rope and better steam engines bigger loads could be lifted up and down the mine shafts.

用钢丝绳和较好的蒸汽机，就能在矿井内上下运输较大的载荷。

6) A construction using wires laid evenly in the helix about a central core has these properties and is able to yield under stress, returning to its original form when the load is removed.

围绕鼓芯均匀地沿螺旋线编捻钢丝的结构就具有这种特性，它在受力时发生变形，而在负载除去之后又恢复其原始状态。

7) For higher pressure multi-stage fans are necessary.

要想得到更高的压力就要采用多级通风机。

8) When the temperature in the fluid coupling exceeds 140 ℃ the alloy plunger will melt and the fluid ejects, leaving the fluid coupling rotating without load.

当液力联轴节中的温度超过 140 摄氏度时，易熔塞即熔化，液体就喷射出来，从而使液力联轴节无负荷空转。

9) Many metal mines and some small coal mines, are ventilated by natural draft alone, which also acts in conjunction with fan pressure in mechanically-ventilated mines, where its importance largely depends on depth of workings and mine resistance.

不少金属矿，以及一些小煤矿，只是采用自然风流通风，但它也和采用机械通风矿井的通风机风压一起起作用；其所起作用的重要程度，大多取决于开采深度和矿井通风阻力。

10) If conditions of humidity and air temperature are favorable, a decided cooling effect on the men is secured by giving the proper velocity to the air current, and the efficiency of the miners is increased.

如果井下空气温度和湿度适当，风流能够保持恰当的速度，便能使矿工明显感觉凉快，从而使矿工工作效率提高。

(二) 用增词法表达出原文的单复数概念

英语中没有量词，而汉语中的量词却往往不可缺少。英语通常依靠冠词和词尾的变化形式表示名词的单复数，但汉语缺乏这种词形变化。鉴于两种语言的这种差异，英汉互译中经常要求适当增加量词、冠词或表示单复数意义的词语。例如：

1) The advantage of a permanent plant is that it is adequate for the conditions likely to arise, but it delays the beginning of sinking and calls for a heavy expenditure at first.

永久性凿井设备的优点是能充分适应可能出现的各种情况，但却要延迟凿井的开始时间，而且需要投入大量的初始费用。

2）Concrete is rammed tightly in between the reinforcements to fill completely the space between the shuttering and the sides of the excavation.

混凝土在钢筋之间振捣实，并填满模板与巷道帮壁间的空隙。

3）Increasing use is made of the bolting method in the coal mines of our country.

锚杆支护法在我国各煤矿得到了日益广泛的应用。

4）When the longwall faceline has reached the designed termination point on the main entries side, a recovery room or entry is established from which all face equipment is recovered, moved to and set up at the setup room for the second panel.

当长壁工作面位置达到主平巷一侧的设计终端点时，要建立一个回收拆卸硐室（或巷道），所有工作面设备要在这里回收拆卸，并搬运到第二个盘区的安装硐室，并进行安装。

5）It has the disadvantage that there are limits to production because it is cyclic mining, that is, it involves separate operations, which includes cutting the coal, drilling the shot holes, charging and shooting the holes, loading the broken coat and installing roof supports, as enumerated above.

普采法的缺点是由于循环开采，产量受到限制，它包括多项独立进行的作业，诸如以上所列举的截煤、钻眼、装药、爆破、装运爆破下来的煤，以及进行顶板支护等。

6）If you look carefully at the outside of the atom you will see tiny bits of matter whirling around the nucleus much the way planets spin around the sun.

如果你仔细观察一下原子的外部，就会看到许多微小的粒子围绕着原子核旋转，其方式几乎和行星围绕太阳旋转一样。

7）The miners are taken to and from the pit-bottoms in cages.

矿工们乘罐笼上下井。

8）Based on the mounting relation between the shearer and the face conveyor there are also two types: the regular type which rides on the conveyor and the in-web shearer which moves on the floor in front of the conveyor.

根据采煤机与工作面输送机的安置关系，采煤机也可分为两类：一类是在输送机上运行的普通型号采煤机；另一类是在输送机前方底板上运行的爬底板滚筒采煤机。

9) The use of booster fans underground is confined to cases where the workings have extended to such great distances from the pit-bottom that the surface fan is incapable of circulating the quantity of air necessary for the ventilation of these remote workings and where it would be necessary either to install a larger and more powerful surface fan or to enlarge the roadways or provide additional airways to allow adequate ventilation.

井下使用辅助通风机仅限于以下两种情况：第一，当巷道离井底车场很远，以致井上通风机无力提供必需的流通风量来为这些偏远的巷道通风；第二，需要安装更大型号、功率更大的井上通风机，或者要拓宽巷道，或者要增加辅助风巷，以便使足够的风量通过。

（三）增添使原意更为确切的词：动词、名词、形容词或副词、介词

1. 增加动词

1) Near the coal-face the roof is lower and the miners may have to bend down as they walk. The only lights are those on their helmets.

在采场顶板较低时，矿工走路可能不得不弯腰。只能用工作帽上的矿灯照明。

2) This is the double-ended ranging-drum shearer with a wheel-rack track haulage system.

这是双端双滚筒可调高采煤机，它采用轮架牵引系统。

3) On the other hand, if both drums are located at one side of the shearer, the face end in the opposite direction still needs a niche.

另一方面，如果两个滚筒都安装在采煤机的一侧，则相反方向的工作面端头部仍然需要开缺口。

4) The large capacity shearers are generally equipped with two electric motors: one for the haulage unit and one gearhead and the other for the other gearhead and other ancillary equipment.

大功率滚筒式采煤机一般装有两台电动机：一台带动牵引部及一传动机头；另一台带动另一传动机头以及其他辅助设备。

5) Water coming into the shaft must be removed by pumping, or, if in too large a quantity for pumping, it may be cemented off.

进入井筒的水必须用泵排出，如果涌水量过大，也可以用水泥灌浆封堵。

2. 增加名词

1) Every precaution is taken to keep the holes vertical if possible, so that each will control the same volume of ground around it, thus ensuring uniform

freezing.

如果可能的话，要采取一切预防措施使钻孔保持垂直，以便使每个钻孔在其周围地层中能拥有相同体积，从而保证达到均衡冻结。

2) For this type of roof, it is necessary to perform proper induced cavings in terms of time and operation.

对于这种类型的顶板，必须根据时间和作业要求，实行恰当的人工强制放顶。

3) The stone removed from the roof to make the roadway high enough is packed, to give support, along gob side of the roadway.

为使巷道保持足够高度而从顶板上取下的石块，沿巷道采空区一侧码成矸石垛，可起支护作用。

4) Some of the most costly sinking in the past are those to be sunk through loose, running, water-bearing sand and weak rocks which always increase the cost enormously.

过去凿井中耗财最多的是那些需要穿过疏松、流动的含水砂层和弱岩层开凿的工程，因为穿过这些岩层总要增加很多费用。

5) Since a powered support consists of four major components (i. e. canopy, caving shield, hydraulic legs or props, and bass plate), the ways by which they are interrelated are used for classification.

液压支架由四种主要部件组成(即：顶梁、掩护梁、液压支柱和底座)，因此，这四种主要部件之间的相互联系方式可以作为液压支架的分类方法。

3. 增加副词或形容词

1) If it is necessary to increase the mining height of the shearer, the drum diameter has to be changed and if necessary, the body height (or the height of the underframe), the length of the ranging arm, and the swing angle have also to be changed.

如需增大采煤机的采煤高度，就必须改变滚筒的直径，而且如果必要，机身高度(即机身底托架高度)、摇臂长度以及回转角度都必须进行相应的改变。

2) Sectional operation is suitable for drill and blasting and plough face.

分段作业较适合炮采和刨煤机工作面。

3) One is with one drum mounted on each side of the shearer's body.

一种是采煤机两侧各安装一个滚筒。

4) The armored face chain conveyor has a large carrying capacity.

铠装工作面刮板输送机的搬运能力非常强大。

5) The resulting plant debris, called peat, accumulated until in some places it was many feet thick.

最后形成的植物残渣，所谓泥煤，不断聚集起来，直到在一些地方达到几英尺厚。

6) 结合理论计算与现场经验，31509 工作面从垂深 25 米处开始探测，32509 工作面从垂深 55 米处开始探测。

Based on the theoretical results and field experience, the initial observation points at faces 31509 and 32509 were located at positions with vertical depths of 25 and 55 m, respectively.

4. 增加介词

1) 高延法提出了采动覆岩的破裂带、离层带、弯曲带和松散冲积层带四带模式。

Gao used a four-zone model, with a failure zone, separation zone, bending zone, and loose alluvium zone.

2) 近距离煤层开采后，覆岩受到反复开采扰动，其导水裂隙不同于单一煤层。

HCF development is different when mining coal seams that are near each other, due to the repeated disturbance of the overburden.

3) 微山湖矿区是中国最大的自然湖泊压煤矿区，湖下压煤量近 7 亿吨。

The Weishan coal area is the largest mining area under a lake in China, with nearly 700 million tons of coal reserves lying under the lake.

4) 待测试孔段的注水流量与孔壁裂隙的漏水流量稳定后，通过流量仪表测定单位时间内的注水量，即孔壁的漏水量。

When the water injection and the leakage from the fractures in the wall become stable, use a flow meter to measure the injection rate, in other words, the rate of water leakage from the wall.

5) 图 8 为 31509 和 32509 工作面开采后覆岩裂隙分布状况。

Figure 8 illustrates the distribution of fractures in the overburden after completion of mining at working faces 31509 and 32509.

(四) 逻辑性增词

英语的逻辑性往往体现在语法和上下文语境中，汉语则暗含在文本之中，因此，翻译时要作适当的增添。例如：

1) Using this method, there is no need to sump the coal back and forth, thus reducing the face-end operations.

使用这种方法，无须往返进刀，从而简化了采煤工作面端的作业工序。

2）As the coal face advances each day, the rocks forming the roof tend to sag down and fill up the "waste", that is, the space where the coal was.

随着采煤工作面每日向前推进，形成顶板的岩石不断沉降以至填满采空区，即填满原来是煤的空间。

3）The main tunnels are lit by electric lights and so is the coal-face.

井下主巷道用电灯照明，采煤工作面也采用电灯照明。

4）The new techniques, including the millisecond blasting, the individual hydraulic props and the large power chine conveyors, have been spread and applied in the drilling and blasting faces.

微差爆破、单体液压支柱和大功率输送机三种新技术，已推广和应用在钻孔和爆破工作面中。

5）根据裂隙发育程度，可自下而上将采动覆岩分为裂隙贯通带、裂隙发育带、微裂隙带和未扰动带。

Based on the level of fracture development, the overburden can be partitioned into four zones: the interconnected fracture zone, fracture development zone, microfracture zone, and unfractured zone.

6）超声波成像、钻孔 CT 及瞬变电磁法技术要求高、程序复杂且误差大，一般作为探测的辅助手段。

Ultrasound imaging, CT of boreholes, and transient electromagnetics are usually used as supplementary methods because of their strict requirements, complex operations, and large errors.

“超声波成像、钻孔 CT 及瞬变电磁法技术要求高、程序复杂且误差大”为后一句“一般作为探测的辅助手段”的原因，所以译文中增加 because of 符合上下文逻辑关系，同时也符合英文表达习惯。

（五）翻译中增加必要的代词

1）It is possible to balance these costs to give the most economical size so far as ventilation is concerned, but the size must also adequate for coal-winding purpose.

就通风而言，我们可以通过计算这些费用来定出最经济的井筒断面尺寸，但是，这种尺寸首先必须保证提煤的需要。

2）So far as most energy supplies are concerned, they are in the form of fuels such as coal oil, where energy is stored as internal energy.

就大多数能源来说，它们都是以燃料的形式（如煤油）存在的，能量是作为内能贮存在这种形式中的。

3) The silicon combines with dissolved oxygen in the steel, thereby improving the quality.

硅能与钢中的溶解氧化合,这就改善了钢的性能。

4) It is believed that these changes are still going on and that given enough time and the proper conditions of temperature and pressure, present-day soft coal might become anthracite or perhaps even graphite.

人们认为,这些变化仍在进行,只要有足够的时间和合适的温度和压力条件,今天的烟煤以后也会变成无烟煤,甚至变成石墨。

5) 在此基础上,分析了对地表湖水进行原位保护的可行性,提出了防治采场顶板砂岩突水的关键技术。

We then analyzed the feasibility of in situ protection of the surface water and proposed techniques to prevent the water in the overlying sandstone from rushing into the work area.

第二节 省略法

省略法,是指在不减少原文词汇所表达的实际概念、不影响原文的思想内容的情况下在译文中省去原文中多余的词。所省译的词语往往是原文结构视为必不可少的,而译文结构视为累赘的词语。英译汉中主要省去一些可有可无、不符合译文习惯表达的词语,如实词中代词、动词的省略;虚词中冠词、介词和连词的省略等等。汉译英主要省略冗词赘语,以及一些表示概念范畴的词语和过分详细的描述。

一、省译英语结构上必需,但汉语语法上不必要的词

1. 省略代词

英语的代词使用频繁,而汉语使用频率较低,英译汉时需要省略大量的代词。例如:

1) The miner has a lamp fixed to his helmet so that he can see to work. The lamp is lit by a battery which is fixed on the miner's belt.

矿工把矿灯固定在工作帽上,从而能看清楚工作场所。矿灯由挂在腰带上的蓄电池点亮。

2) The feature of gas accumulation is that gas content in east coal-field is higher than that in west zone and also gradually increases from shallow to

deep.

瓦斯积聚特点是东部煤田瓦斯含量高于西部，并由浅到深逐渐增加。

3）The coal-bearing area of Pingdingshan Coal (Group) is 2 374 km^2 and it is the production base of coking and power coal with the most complete product variety domestically.

平煤集团矿区面积为 2 374 平方千米，是国内煤炭种类齐全的焦煤、电煤生产基地。

4）The modern longwall face, advances so rapidly that it imposes two requirements on the entry transportation: it must have a high carrying capacity and it must be able to move along as the face advances.

现代化的长壁采煤工作面推进速度很快，这就给顺槽运输提出了两个要求：必须具有很大的运载能力，并能随着工作面的推进而前移。

5）The basic principle of bolting is that it should make the bolted rock an integral part of supporting structure.

锚杆支护的基本原理是使锚杆连接的岩石成为支承结构的一个组成部分。

2. 省略冠词

英语中，冠词用得很广泛，汉语中却没有冠词。一般情况下，英语不定冠词 a、an 如果不具有数量词 one 的含义便可省去；英语定冠词 the 如果不具有指代词 this、that、these、those 的含义，也可省略。例如：

1）The tool (that) he is working with is called a wrench.

他干活用的那种工具叫作扳手。

2）The stationary pump and motor is mounted on the same bed-plate and the entire unit may be loaded on a truck for transportation.

固定式水泵和马达组装在同一底座上，整个装置也可以装载到车上以便运输。

3）One continuous miner can mechanically break apart about 1.8 tons of coal per hour.

连续采煤机可机械落矿，每小时约 1.8 吨煤炭。

4）An inclined shaft has been developed, one belt conveyor (belt width is 1 000 mm) erected and hoisting changed from the original skip hoisting to belt conveying.

开凿了斜井，安装了带式输送机（带宽 1 000 毫米），由原来的箕斗提升改为带式输送机输送。

5）As soon as one hole cuts a fissure, the rod is withdrawn and the pipe connected to the cementation pump.

一旦钻孔穿过缝隙,便拔出钻杆,并立刻将注浆管连上注浆泵。

3. 省略介词

大量频繁使用介词是英语的特点之一,而汉语中介词使用的频率要低得多;汉语句子成分之间的关系往往由词序和逻辑关系体现出来。因此,英汉翻译时,须省略可有可无的介词,以及介词短语。例如:

1) There are two gearheads: one on the left, and the other on the right-hand side of the shearer.

共有两个传动头,左右各一个。

2) The number and spacing of the blades per stage varies, being usually about 12 to 16.

叶片数和叶片间距因级而异,通常约在12～16片之间。

3) The armored face chain conveyor is structurally strong, bendable like a snake and low in body height.

铠装工作面刮板输送机的结构牢固,能弯曲如蛇,而且机身低矮。

4) The total thickness of coal measures strata in the north limb of the coal-field is 610 m. It has about 50 coal seams, the total thickness of coal seams is 29 m and the coal-bearing coefficient is 4.75 percent.

煤田北翼岩层厚度610米,大约50层煤,煤层总厚度29米,煤层含煤系数为4.75%。

5) Coal dust of all coal seams in the coal-field has explosion hazard and the ignition period is 2 to 12 months.

煤田所有煤层的煤尘都具爆炸危险,易爆期为2～12个月。

4. 省略连词

汉语词语之间连接词用得不多,其上下逻辑关系常常是暗含的,由词语的次序来表示。英语则不然,连接词用得比较多。因此英译汉时在很多情况下可不必把连接词译出来。例如:

1) Steel support can be erected in a shorter time and by fewer men than timbering, and, especially, due to its good supporting properties, the trend toward use of steel has been encouraged.

钢材支架的架设,比木材支架省时省力,特别是由于钢材支架具有良好的支护性能,因而目前正被鼓励大量使用。

2) The weight of sediments compacted the peat, which during the ensuing millennia became more dense and was gradually converted to coal.

沉积物的重量紧压着泥煤，在长达数百万年的过程中泥煤变得日益稠密，从而逐步形成了煤炭。

3）The action of the axial-flow fan differs from that of the centrifugal fan in that the air passes axially along the fan instead of being discharged from the circumference of the fan by centrifugal force.

轴流式通风机的工作原理与离心式通风机不同，其风流不是通过离心力作用从通风机周边排放，而是沿通风机轴向流过。

4）Security is essential to life while violation is the source of illegal accident.

安全是生命之本，违章是事故之源。

5）The winding ropes are attached one at each end of the drum barrels and arranged to coil on the drum in opposite directions, so that when the drum rotates one cage will be raised and the other lowered.

两根提升钢丝绳各系结在滚筒的一侧，并且缠绕方向相反，滚筒旋转时，使一个罐笼上升，另一个下降。

5. 省略动词

英语句子严格遵循主谓结构，而汉语句子有时没有谓语。当英语句子中的动词不表示实际意义，只是表达状态或是作比喻性或习惯性用法时，汉译时可以省略。例如：

1）Girders carried between the walls may be used to support the various stagings required for the heapstead structure.

插在井壁结构中的一些横梁，可用来支撑井口建筑的各种构架。

2）The bit produced by new technology has high strength, good wearability and less consumption.

采用新工艺生产的截齿，其强度高、耐磨性强、损耗小。

3）Volute centrifugals are so called because the casting is of the volute type (similar in design to a ventilating fan casting).

螺旋离心泵之得名是因为壳体呈螺旋形(构造很像一个通风机的外壳)。

4）Common drilling method is simple, easy to master, safe and reliable.

普通凿井法操作简单、易于掌握、安全可靠。

5）浅孔爆破多用于井巷工程；中深孔爆破多用于井筒及大断面硐室掘进；深孔爆破主要用于立井井筒及溜煤眼、大断面硐室以及露天开采的台阶爆破。

Shallow depth blasting is used for roadway construction; medium-depth blasting for large cross-section shaft and cavern room boring; deep-hole blasting mainly for the shaft and coal chute, a large section of the chamber and the open pit bench blasting.

英文中同样的结构重复出现时省略,此译例中省略了反复出现的动词 used。

6. 省略无实际意义的语助词 it 和 there

1) There are, however, certain drawbacks associated with longwall retreating, the chief of which are the great initial cost, involved in driving out the headings to the boundary and the long period of time which elapses before the mine reaches full production.

然而,后退式长壁开采法也有一定的缺点,其中主要是初期费用较高,这包括需要将煤巷掘到边界,同时需要很长时间才能使矿井达到设计产量。

2) The application of modern powered supports can be traced back to the early 1950s. Since then, following its adoption in every part of the world, there have been countless models designed and manufactured in various countries.

现代化液压支架的使用,可以追溯到 20 世纪 50 年代早期。从那时起,随着世界各地的采用,各个国家都设计和制造了无数不同型号的液压支架。

3) However, it must be noted that the trend of development in each type is such that it becomes less distinguishable in terms of application.

然而,必须指出的是各种类型液压支架的发展趋势使它们在使用方面的区别变得越来越小。

4) This is the first evidence of coal being deliberately dug from the ground, but it is quite possible that even before this, coal was used for fuel in some parts of the world.

这是人类有意识地从地下挖煤的最早证据,但是很可能甚至在这以前,世界上有些地方已经用煤作燃料了。

5) Thus it is possible to ignore most of the problems introduced by the curvature of the earth and by the convergence of the meridians, except in the most precise work.

因此,除了最精确的工作以外,忽略由于地壳曲率以及经线集中引起的问题是可行的。

二、出于逻辑上的考虑省译多余的词语

在英汉两种语言中,往往会出现一些不言而喻的多余词,这些多余词从译入

语逻辑角度看纯属多此一举，如果一字不漏地翻译，译文不仅有翻译腔，而且有悖事理。因此，翻译时可省译多余词。例如：

1) The total thickness of main minable seams is 15-18 m and 6 seams (G21, F14, E11, D7, D4 and B2) are partly minable.

主要可采煤层有15～18米厚，G21、F14、E11、D7、D4和B2六个煤层为部分可采煤层。

the total thickness of main minable seams 如果译为“主要可采煤层有15～18米总厚度”，行文不符合汉语表达习惯，故省译 total。

2) Also included would be blocks of rock which are broken by fracture or joint patterns in such a manner that they may subsequently loosen and fall.

应当撬落的还有那些被裂缝或节理纹路切割要松动和垮落的岩块。

subsequently 表达的含义已经在译文的字里行间，故省译。

3) As sinking proceeds, this temporary lining is removed and replaced by permanent lining, which may usually be lined solidly with concrete to shut water out of the shaft and render the shaft fireproof.

随着凿井工序向前推进，原来的临时井壁应拆除，并代之以永久性井壁，永久性井壁通常可以混凝土整体砌筑，以防止地下水进入井筒并使井筒能够防火。

4) The leakage curves for boreholes 4# and 5# intersect at a vertical depth of 85.1 m, and those for boreholes 4# and 6# meet at a vertical depth of 88.5 m.

4#和5#钻孔漏失量曲线相交于垂深85.1米处，4#和6#钻孔漏失量曲线相交于88.5米处。

第二个 a vertical depth of 为重复表达，故省译。

5) After reaching the frozen thickness designed, we can assemble the shaft excavation operations until going smoothly through the unstable ground.

冻结壁达设计厚度后，即可进行井筒掘砌作业，直到顺利穿过不稳定地层为止。

三、出于修辞和表达习惯的考虑省略多余的词语

1) The decision on the development of open pit mine substantially depends upon the stripping ration which is defined as the weight of surface soil and cover rock to be removed for extraction of unit weight of mineral ore.

露天矿开拓的决策很大程度上取决于剥离率，它被定义为开采单位质量矿石剥离的表土和上覆岩层质量。

在 to be removed for extraction 中，remove 和 extraction 均为“移除，剥离”的意思，译文中二者含义重复，因此翻译时省译 to be removed。

2) It is questionable whether it will prove to be practicable for great depths, largely because of the difficulty of maintaining verticality of the boreholes for the freezing pipes.

它是否适用于深度挖掘，尚未确定，主要是因为在冷冻管道中很难保持钻孔的垂直。

3) When the fall height of the falling strata is more than 2.5 times the thickness of the falling strata, the strata will fall rapidly and rotate during falling, resulting in irregular rock fragments on the floor.

当垮落岩层的垮落高度大于其厚度 2.5 倍时，在垮落过程中，岩块将急速下落并出现翻转现象，在底板上形成规则的碎石堆。

句中有两个 the falling strata，为了避免重复，第二个 the falling strata 省译，用“其”来替代。

4) 综合考虑 2 个采后钻孔的探测结果，确定 31509 工作面导水裂隙带高度为 60.8 米。

The results from the two post-mining observation boreholes reveals that the height of the HCF zone at face 31509 was 60.8 m.

5) 为减少钻孔斜长，降低工程费用，并利于钻孔内的碎屑及渣石排出，确保堵水器的封堵效果。在施工条件允许的情况下，应尽可能增大钻孔倾角。

To reduce drilling length and engineering cost, increase effective flushing of the boreholes while drilling, and ensure effective of plugging to create a sealed borehole interval, the borehole inclination should be as great as possible.

根据英文表达习惯及上下文，译文中省略冗余表达“在施工条件允许的情况下”。

第三节 词类转换法

由于英汉两种语言分属两个不同的语系，它们在语言的表达方式上各有各的特点。尽管它们在词类划分上大致相同，但词类的使用频率在两种语言中却有差异。英语中频繁使用名词和介词，而汉语中经常使用动词。因此，在英汉互译中，经常要用到词类转换法，即根据上下文和译文的表达习惯，在不改变原文词义的前提下，将原文中某些词的词性在译文中作相应的改变，使译文读起来更加通顺。

一、汉语动词和英语词类的转换

英语与汉语比较起来，汉语中动词用得比较多。英语句子中往往只能有一个谓语动词，而在汉语句子中却可以几个动词或动词结构连用。例如：确定漏失规律和揭示漏失机理，探索和研制防漏堵漏方案及工艺，对提高钻探工程质量、增加钻探工程效益具有现实和长远意义。这一个汉语句子中，就有七个动词。翻译成英语时，只能保留一个主要的动词“具有”作谓语动词，其他的需要转译为英语的非谓语动词。可见，英语中不少词类可以和汉语的动词互相转换。

1. 英语名词转变成汉语动词

英语学术语篇广泛使用名词化结构，即用动词的名词化形式来表示动作意义，往往后面跟 of 介词短语，如 the rotation of the earth on its own axis。名词化结构也可以表示时间、原因、目的等状语或状语从句。名词化结构表达客观存在，把动态的动作抽象为事实，传递信息，实现抽象、技术性和客观公正的效果，也可使复合句变为简单句，使表达的概念更加确切严密。由于汉语词语缺乏形态变化，所以学术语篇以动词结构为主。英译汉时，我们需要把这些抽象的名词化结构转变成汉语的动词。例如：

1) *Observation* on the rock surface revealed that the lines in the rock are curved and discontinuous.

观察岩层表面，发现岩体纹路通道是弯曲以及不连续的。

2) The *comparison* between prediction and test results for complex stress path reflects that the model's stress path of the stress-strain is reasonable.

通过复杂应力路径的预测和测试结果的对比，可认为模型应力-应变曲线的应力路径是合理的。

3) An abandoned coal mine is a closed-up mine caused by the depletion of coal resources, *integration* of resources, *eliminating* of outdated production capacity and other factors.

废弃矿井是由于煤炭资源枯竭、资源整合以及淘汰落后产能等因素造成的已关闭的矿井。

4) Proper *selection* of circuit components permits a transistor to operate in this characteristic region.

正确地选择电路元件能使晶体管在这个特殊领域工作。

5) The first step in the *solution* of any dynamic problem is the selection of an appropriate coordinate system.

在解任何一个动力学运算题时，要做的第一步就是选择一个合适的坐标。

6) One major cause of unsafe practices is *lack* of knowledge or skill.

不安全操作的一个重要原因是缺乏知识或技能。

例1)到例6)的7个英语名词(斜体)均为用动词的名词化结构来表达动作的意义，因为它们要在句中作主语、宾语或表语，所以都发生了词形或词性的变化，翻译成汉语时都还原成了动词。

7) *Neglect of environment in the course of industrialization*, particularly the irrational exploitation and utilization of natural resources, has caused global environmental pollution and ecological degradation, posing a real threat to the survival and development of mankind.

由于工业化过程中不重视环境，尤其是不合理地开发利用自然资源，造成了全球性的环境污染和生态破坏，对人类的生存和发展构成了现实威胁。

8) *Seismic measurements of travel time and amplitude* define the subsurface geometry and give an estimate of the acoustic impedances related to rock velocities and densities.

如果我们测得地震波的走时和振幅，我们就能够确定地下的几何形状并估算出岩石速度和密度有关的声阻抗。

9) *Substitution of the actual values into the equation* results in $a=b$.

把实际数值代入该方程式后，就会得出 a 等于 b 的结果。

10) *The raise of efficiency of utilizing energy* depends on the scientific and technological progress in energy, power, petroleum and chemical industries.

要提高能源利用效率，也就是要依靠能源、动力、石油、化工等行业的科技进步来满足。

例7)到例10)的四个名词化结构，表示的分别是原因、条件、时间和目的状语从句的意义，把复合句变成了简单句，达到了言简意赅的效果。翻译成汉语时，这些名词化结构都翻译成了汉语的动词。

2. 英语介词转变成汉语动词

英语介词在组句时比汉语介词活跃、功能性更强、使用更为广泛。有些介词本身就是由动词演变而来的，或者具有动作的意义，如 with、against、past、toward 等，因此英译汉时，常常转换为汉语的动词，以符合汉语的表达习惯。例如：

1) Cabentonite (clay) is a special mineral *with* suspension, expansion and fire-resistant properties.

膨润土(陶土)是一种具有悬浮性、膨胀性和耐火性的特殊矿物。

2) Rock is *of* heterogeneity and discontinuity, and research on its dynamic fracture property has been one of the hotspots and challenges in domestic and foreign studies.

岩石具有非均质性、不连续性,其动态断裂性能研究是国内外研究的热点和难点问题之一。

3) A force is needed to move an object *against* inertia.

为使物体克服惯性而运动,就需要一个力。

4) *With* a mind toward practical applications, dozens of labs are working on wires and films thin enough to deposit on computer chips.

由于对超导体的实际用途都抱有希望,几十家实验室都在研制可以固定在计算机芯片上的细线和薄膜。

5) Strip mining is an effective method *for* mining without destroying the aquifers. The key parameter is the ratio of mined to unmined width, which is designed based on real hydrogeology conditions.

条带采煤法是目前实现保水开采的一种行之有效的方法,关键技术是根据具体的水文地质条件科学设计开采煤层的采留比参数。

6) Coal of Group B has high heating value *with* low ash, phosphorus and sulphur.

乙组煤属低灰、低磷、低硫高热值煤。

3. 英语形容词转变为汉语动词

英语中有些形容词含有动词的意义,尤其是以-ble 为结尾的词或者和介词搭配使用的形容词;还有一些形容词是由相关动词派生出来的。翻译时,这类形容词需要转译成汉语的动词。例如:

1) Heat is a form of energy into which all other forms are *convertible*.

热是能量的一种形式,其他一切形式的能量都能改变为热能。

2) The two bodies are so far apart that the attractive force between them is *negligible*.

这两个物体相距如此之远,它们之间的吸引力可以忽略。

3) Water is a substance *suitable* for preparation of hydrogen and oxygen.

水是适于制取氢和氧的物质。

4) If low-cost power becomes *available* from nuclear plants, the electricity crisis would be solved.

如果能从核电站获得低成本的电力,电力紧张问题就能解决。

5) Lime can be obtained from the calcination of limestones *present* widely in nature.

自然界中广泛存在石灰石，人们可以通过煅烧石灰石来提取石灰。

另外，英语中一些表示情感、知觉、愿望、态度等心理状态的形容词常跟在系动词后面构成复合谓语，结构为“be/get/其他系动词＋形容词＋介词短语或从句”。这些形容词在译成汉语时，也需要转译为汉语动词。例如：

6) Simple, low-cost colorimeters and spectrophotometers *are* usually *adequate for* relative color measurements.

功能简单、价格便宜的色度计和光谱仪通常能满足相应的色度测量。

4. 汉语动词转变成英语名词、形容词或介词

1) 由于线圈中*存在*铁，磁感应强度增加到原来的 5 500 倍以上。

The presence of the iron in the coil has increased the magnetic induction to over 5 500 times what it was.

2) 应当始终*注意*保护仪器，不使其沾染灰尘和受潮。

Care must be taken at all times to protect the instrument from dust and damp.

3) 科学家们*深信*所有的物质都是不可毁坏的。

Scientists are confident that all matter is indestructible.

4) 水在 4 ℃以下就不断地*膨胀*而不是不断地*收缩*。

Below 4 ℃, water is in continuous expansion instead of continuous contraction.

5) 这个项目现在正*处于*初级阶段，预期在 2027 年开工。

The project is in its early development phase and is expected to start in 2027.

6) 在分析潞安矿区地质条件、钻探生产工艺、钻孔堵漏研究现状基础上，*探索*并*划分*了潞安矿区钻孔漏失类型，*调查*并*研究*了漏失地层特征与分布规律，分析并确定了漳村井田煤田钻孔漏失原因。

Based on the analysis for the geological conditions of Lu'an mining area, the drilling production technology and the present situation of plugging, the exploration and division for the circulation loss mode of Lu'an mining area, as well as the investigation and research on the leaking formation pattern and distribution were carried out to study and determine the reasons for the drilling circulation loss of Zhangcun coal mine.

以上例句中，例 1)、例 2)和例 6)中几个汉语动词(斜体)均要在英语译文中

作主语，所以都由动词转变成了名词，用名词化结构把动态的动作抽象为事实来传递信息。例 3)中的“深信”表达人的心理状态，在英语中习惯用形容词或形容词词组，所以做了汉语动词到英语形容词的转换。例 4)和例 5)中的汉语动词都是表达的物体所处的状态或阶段，所以均转换成了英语介词 in。

二、汉语名词与英语词类的转换

1. 英语动词转变成汉语名词

这种翻译方法一般用于汉语中只能用名词来表达的一些概念，这些概念往往没有相应的动词表述。而在英语中表达这些概念的词，通常既可以作名词，又可以作动词。例如：

1) The design *aims* at automatic operation, easy regulation, simple maintenance and high productivity.

设计的目的在于自动操作、调节方便、维护简易、生产高效。

2) Shields are *characterized* by the addition of a caving shield at the rear end between the base and the canopy.

掩护式支架的特点是在底座与顶梁之间的后部增加了一个掩护梁。

3) During the process of impact, the position of maximum tension in bolt *tends* to move to the middle of bolt.

在挤压过程中，锚杆最大拉力的位置有向锚杆中间移动的趋势。

4) If phlogiston left the metal during calcination, the metal should *weigh* less.

如果金属在燃烧时有燃烧素损失的话，它的重量就会变轻。

5) The physical properties *are concerned with* the characteristics of coal in its natural state, or prior to its end use as a fuel.

煤的物理性质与其在自然状态或作为燃料最终使用之前的特性有关。

2. 英语形容词转变成汉语名词

说明事物的形状时，英语往往习惯用表示特征的形容词及其比较级作表语表达，汉语却常在其后添加“性”“度”等缀词，习惯用名词来表达，如“强度”“硬度”“密度”“弹性”“流动性”等。因此，英译汉时通常要作从形容词到名词的转换。

1) Glass is much more *soluble* than quartz.

玻璃的可溶性比石英大得多。

2) The circulation loss is one of the *multiple* and *cataclysmic* accidents in drilling engineering.

钻孔漏失是钻探工程多发性和灾难性事故之一。

3) In the fission processes the fission fragments are very *radioactive*.

在裂变过程中，裂变碎片具有强烈的放射性。

4)The cutting tool must be *strong*, *tough*, *hard* and *wear-resistant*.

刀具必须有足够的强度、韧性、硬度，而且要耐磨。

5) Fuel leaks are extremely dangerous as gasoline is very *flammable*.

汽油的可燃性很强，所以汽油泄漏极其危险。

3. 英语副词转变成汉语名词

由于表达习惯上的差异，有些英语副词（主要是一些以名词作词根的副词）虽然在句子中作次要成分（状语），但其表达的意义和概念却在句中占有重要地位；此外，这些副词往往不好译成状语的形式，所以我们会把它们转换成汉语的名词形式。例如：

1) The blueprint must be *dimensionally* and proportionally correct.

蓝图的尺寸和比例必须正确。

2) There seem to be no other competitive techniques which can measure range as *well* and *rapidly* as a laser can.

就测距的精度和速度而论，似乎还没有其他技术可与激光相比。

3) Oxygen is one of the most important elements in the physical world, and it is very active *chemically*.

氧是物质世界最重要的元素之一，其化学性能很活泼。

4) All structural materials behave *plastically* above their elastic range.

超过弹性极限时，一切结构都会显示出塑性。

5) The transition between phases of a substance can be *graphically* displayed.

物质的各种相变过程可以用图表展示。

4. 汉语名词转变成英语动词、形容词或副词

1) 综合机械化采煤的*特点*是滚筒采煤机或刨煤机，综合移动链式运输机和液压支柱的使用。

The mechanized mining is characterized by using the shearer or plow, integrally removable chain conveyor and hydraulic props.

2) 这几种化合物的*沸点*各不相同。

Each of the compounds boils at a different temperature.

3) 动量的*定义*是速度和物体质量的乘积。

Momentum is defined as the product of the velocity and a quantity called the mass of the body.

4）中碳钢的*硬度*比低碳钢的大很多。

Medium carbon steel is much stronger than low carbon steel.

5）但是在多次失败之后，人们发现长壁开采法在美国还是极具*可操作性*的。

After some failures and misapplications, longwall mining has become practical in the USA.

6）节式支架很简单，而且更加灵活，但是*结构*不够稳定。

The frame support is very simple, more flexible but less stable structurally.

上述例子中，例1)到例3)的三个名词“特点”“沸点”“定义”换成了英语动词词组 be characterized by、boil at 以及 be defined as，这样更符合英语的表达习惯。例4)和例5)中的两个名词则转变成了英语的形容词或形容词的比较级。例6)则转译成了英语的副词。

三、汉语形容词与英语词类的转换

1. 英语副词转变成汉语形容词

英语的动词和形容词可以转变成汉语的名词，因此修饰动词和形容词的副词也可以随之转译成汉语的形容词。例如：

1）Earthquakes are *closely* related to faulting.

地震和断裂运动有密切关系。

2）The universal lathe is most *widely* used and plays an important part in industry.

万能机床得到了广泛的应用，因而在工业上起着重要的作用。

3）The application results show that the super-high-water material is a gob backfilling material that performs *well* in a wet, cold and closed underground environment.

应用结果表明，超高水材料是一种在潮湿、低温和封闭的地下环境中具有良好性能的采空区充填材料。

4）However, Northwest China is *geologically* characterized by dry climate, sparse vegetation and short water resources accounting for only 3.9% of the country's total.

然而，西北地区的地理特点是气候干旱、植被覆盖度低、水资源短缺，仅占全国水资源的3.9%。

5) The pressure of gas is *inversely* proportional to its volume, if its temperature is kept constant.

如果温度保持不变,则气体压力与其体积成反比。

在上面前四个例子中,动词 relate、use、perform 和 characterize 均译成了汉语的名词,因此修饰它们的副词 closely、widely、well 以及 geologically 也随之转译成相应的形容词。例 5)里,形容词 proportional 翻译成了汉语的名词"比例",所以修饰形容词的副词 inversely 也相应地转译成了汉语的形容词"相反的"。

2. 英语名词转变成汉语形容词

英语中的某些名词,特别是一些形容词派生的名词,作表语或宾语时,汉译将其转换成形容词更符合汉语的表达习惯。例如:

1) Single crystals of high perfection are an absolute *necessity* for the fabrication of integrated circuits.

对制造集成电路来说,高纯度的单晶是绝对必要的。

2) The simulation experiment was quite a *success*.

此次模拟实验非常成功。

3) The lower stretches of rivers show considerable *variety*.

河流下游的情况是大不相同的。

4) In a 3D environment, rich feedback in the form of cursor and other types of hinting becomes a *necessity*.

在三维环境中,以光标和其他类型的暗示来显示丰富的视觉反馈是非常必要的。

5) Thermoelectric power sources are used on deep-space probes because of their *reliability* and *convenience*.

热电动力源由于性能可靠且方便,被用于深层空间探测器上。

3. 汉语形容词转变成英语副词或名词

1) *现在的*黄金是从深埋在地下的石英岩里开采出来的。

Today gold is mined from quartz rocks deep below the earth's surface.

原句中,"现在的"是修饰 gold 的定语,此处我们将其转译成副词,作整个句子的时间状语。

2) 这些材料的*主要*特点是绝缘性好、耐磨性强。

Such materials are chiefly characterized by good insulation and high resistance to wear.

原句中,"……的特点"按英语的习惯表达翻译成了词组 be characterized

by，相应地，修饰“特点”的形容词“主要的”也转译成了英语的副词 chiefly。

3）许多用户意识到，当信息存储在工作站或服务器时，信息保密*非常重要*。

Many users realize the importance of confidential information when it is stored on their workstations or servers.

原句中，“非常重要”是形容词作表语，而在英语译文中我们把句子作了整合，让“信息保密非常重要”作“意识到”这个谓语动词的宾语，所以我们将其转译成名词 importance，作整个大句子的宾语。

第四节　反译法

由于生活方式和思维习惯的差异导致英语和汉语存在很大差异，例如，在一种语言中是肯定的表达在另一种语言中则是否定的表达，反之亦然。

A：Are you not going tomorrow?（你明天不去吗?）

B：No，I'm not going.（是的，我不去。）

每一种语言都有其自身独特的否定表达方式，英语和汉语中往往均可用肯定形式或否定形式表达同一概念，即以肯定译肯定、以否定译否定[也就是用汉语里的“不”“非”“无”“没(有)”“未”“否”等译英语中的 no 或 not 以及一些带有否定词缀的词]。但是由于英汉表达习惯的差异，正译法有时并非切实可行。英语中的肯定形式可以甚至必须译成汉语中的否定形式，而英语中的否定形式却宜译作汉语中的肯定形式，这种翻译技巧即“反译法”，也称作“反面着笔法”或“正反、反正译法”。如英语中的标志用语“Wet Paint!”如译作“湿的油漆!”，不符合汉语的表达习惯，而应译成“油漆未干!”。汉语中的货物装箱标志用语“切勿倒置!”译为英语中的肯定表达形式“Keep Upright!”较为合适。

一、英语中否定的表达方式

英语的否定表达方式按语言学家们的划分大致上有四类：

1. 完全否定

完全否定指否定整个句子的全部意思。英语中表示完全否定的单词和词组主要有：no、not、none、never、nothing、nobody、nowhere、neither、nor、not at all。含有这些词的完全否定句一般情况下意思都一目了然，只要注意否定的表达，译文符合现代汉语的习惯即可。例如：

1）The cages never carry men and coal together.

罐笼从来不许同时运送人和煤。

2) The frame support is not suitable for a weak roof.

易受损的顶板不适合使用节式支架。

3) When operating the hydraulic support, the props don't lift.

操作液压支架时不升柱。

4) The fault phenomenon is that the support neither rises nor drops neither pushes nor pulls.

故障现象为支架不升不降,不推不拉。

5) The operation hydraulic support does not drop column.

操作液压支架不降柱。

6) The plow head contains no motorized equipment.

刨头不包含电动设备。

2. 部分否定

如果句中有 all、both、every(包括 everybody、everything 等)、complete (completely)、whole (wholly)、total (totally)、fully、entirely、always、much、many、often、quite、well、altogether 等具有概括意义的代词、形容词和副词与否定词 not 连用时,该句则为部分否定句,翻译时,一般译为“不都”“不全”“不多”“不常”“一些”“并非人人”“并非都”“并不是”等。例如:

1) The main reasons include: ① the manual unloading valve is not fully closed or well sealed; ② the fluid quantity in the tank is insufficient; ③ blocking phenomenon occurs in suction filter; ④ the one-way valve core of the liquid discharging is not well sealed; ⑤ the one-way valve core of the suction valve is not well sealed.

其主要原因包括:① 手动卸载阀没有完全关闭或密封不良;② 液箱内液量不足; ③ 吸液过滤器有阻塞现象;④ 排液单向阀阀芯密封不良;⑤ 吸液单向阀阀芯密封不良。

2) With the axial-flow fan it is possible to vary the performance by increasing its speed, by increasing the number of stages or rotors, and by altering the pitch or inclination of the blades, and these alterations can be made over fairly wide limits while not seriously reducing the efficiency at which the fan works.

在使用轴流式扇风机的过程中,可以通过提高速度、增加级数或转子数,以及变换叶片间距或倾角等方法,使其具有不同性能,而且这些变换可以在相当大的范围内进行而不会严重地降低扇风机的工作效率。

3) With the introduction of machines to expedite and cheapen the cost of headings, however, these disadvantages are not so serious as in the past.

然而，由于采用机器开掘煤巷，降低了成本，缩短了工期，所以这些缺点已不像过去那么严重了。

4) Right after the shearer's cutting if the support is not advanced immediately, the unsupported roof between the faceline and the canopy tip will fall in a short period of time (less than 10 min).

采煤机截割以后，如果支架不立即前移，工作面与顶梁端头之间未支撑的顶板便会很快(十分钟之内)垮落。

3. 半否定

如果句中出现 hardly、scarcely、seldom、barely、few、little 等词时，句子为半否定句。例如：

1) It is rather stable during plowing, with little conveyor rebound.

这种刨煤机在刨煤时相当稳定，几乎不会引起输送机的反弹。

2) There is very little brown coal in England—none at all in coalfields—but very large quantities occur in several other countries.

在英格兰几乎没有褐煤，许多煤田里一点也没有，而在其他一些国家却赋存着大量的褐煤。

3) It is hard and shows little sign of banding, but has a luster (or shines) rather like dull steel, and it breaks into skew-shaped blocks.

这种煤质地坚硬，几乎没有夹层的痕迹，光泽有点像暗色的钢，容易破碎成斜形块。

4. 无否定形式的否定(用带否定意义的词语表达否定含义)

英语中有些词或词组如 absence、fail、few、seldom、little、without、unless、anything but 等，从字面上看并无否定形式，但在表达某种意思时，却具有否定意义。翻译含有这类词或词组的句子时，通常在正确理解意思的基础上，根据汉语习惯，采取正反处理手段，译成带否定意义的句子。例如：

1) Coal was known to man thousands of years ago. Ancient writings tell us that three thousand years ago the Chinese knew that certain kind of black rock would burn, and in one part of the country where there was little wood they used to dig into the earth to find this black rock: for their fires.

几千年以前人类对煤就有所了解。古书告诉我们，在 3 000 年以前中国人就知道某几种黑石头能够燃烧。在这个国家木柴缺少的地方，人们常常挖地寻找这种黑石头用来烧火。

2) A machine fire in an untimbered metal mine airway will remain localized if there is little else to burn in the vicinity.

如果附近没有可燃物，在没有木材的金属矿山巷道中发生的机械火灾将停留在局部地区。

3) In this zone, the strata deform without causing any major cracks cutting through the thickness of the strata as in the fractured zone.

在此(持续变形)带中岩层的变化，不会像断裂带那样造成贯穿整个岩层厚度的大裂缝。

4) In the recent models the shearer loaders are self-propelled along a special track without the aid of the chain.

在最新的机型中，采煤滚筒装载机是沿着一条特殊的轨道自推进，而没有链式运输机协助。

以上否定的翻译要根据具体情况而定，不能机械地套用“正译”或“反译”。

二、英语肯定，汉语否定

英语中有很多表达从形式上来看是肯定的，没有否定词 not、no、never、none 和半否定词 hardly、scarcely、seldom、little、few，但实际上却具有否定意义。翻译时，若从正面翻译是行不通的，这时需要正说反译。这种情况出现于各种表达形式中，包括词(动词、副词、形容词、介词、连词)、短语和句子。

1. 动词

矿业英语中，部分动词，如 restrain、eliminate、vary、reverse 等，在译为汉语时，一般译为否定形式。例如：

1) This requires a construction which is flexible but in which the constituent members are restrained in their relative positions.

这就要求它结构柔软，而其内部组成部分的相对位置始终保持不变。

如果将此句译为“这就要求它结构柔软，而其内部组成部分限定在相对位置”，虽语法上没有错误，但听起来别扭，不符合汉语表达习惯。

2) The system eliminates any types of drive chain.

该系统不用任何形式的传动链。

3) The gearboxes are basically of the same type but vary in design.

基本上变速器的型号是一样的，但外表是不同的。

4) The hydraulic controlled one-way valve is damaged (reverse barrier fluid).

液控单向阀损坏(反向不通液)。

5）With the axial-flow fan it is possible to vary the performance by increasing its speed, by increasing the number of stages or rotors, and by altering the pitch or inclination of the blades.

在使用轴流式扇风机的过程中，可以通过提高速度、增加级数或转子数，以及变换叶片间距或倾角等方法，使其具有不同性能。

2. 副词

常用来表达否定意义的副词有 exclusively、only、less 等。例如：

1）The portable type of pump is used almost exclusively in gathering service and the entire unit is mounted on a truck which makes transportation a simple matter.

轻便型水泵几乎毫无例外地用于汇集水流，也可以将整个装置安装在车上，以便移动。

2）The frame support is very simple, more flexible but less stable structurally.

节式支架很简单，而且更加灵活，但是结构不够稳定。

3）If the seam thickness varies considerably, the thickness of extraction as established before must be implemented uniformly.

如果煤层厚度变化大，应始终保持先前确定的开采厚度不变。

如果将后半句译为“应始终保持先前确定的开采厚度一致”，语句听起来别扭，不符合汉语的表达习惯。

3. 形容词

常用来表达否定意义的形容词包括 different、以-less 为后缀的形容词、various 以及有否定前缀的形容词如 unsafe 等。这些词在英语句中出现，其汉语应译为否定。例如：

1）It is to be noted that exploration has to be deterministic, but the availability of oil and gas is estimated based on probability.

值得注意的是，勘探结论具有不以人意志为转移的必然性，而油气储量估算结果却是概然性的。

2）The electric haulage shearer includes chainless haulage unit, two gearheads, auxiliary power pack, haulage and auxiliary control.

电牵引采煤机包括无链牵引单元、两个齿轮、辅助电源箱、牵引和辅助控制。

chainless 在形式上是肯定的，但含义是否定的，所以翻译的时候应译为否定。

3) The most important item is, based on the characteristics of surface movement and deformation, how to make use of the adaptability of the structure and attempt to eliminate or reduce the effects of the factors that are detrimental to structural stability.

最重要的一条是,根据地表移动和变形的特性,如何利用结构的适应性和如何设法消除或减少那些不利于结构稳定的因素的影响。

4) Hard and fast rules distinguishing between these various methods cannot be laid down in some cases, as so many modifications and combinations of the methods exist at various collieries working under different conditions.

要规定一条严格的准则来区分上述各种开采方法有时是不可能的,这是因为在不同的条件下开采的各个煤矿采用了许多改进的和综合的开采方法。

5) One major cause of unsafe practices is lack of knowledge or skill.

不安全操作的一个重要原因是缺乏知识或技能。

4. 介词

部分介词,如 without、unlike 等在英语句中出现,其汉语应译为否定。例如:

1) In practice, therefore, it is usual to restrict the ratio of diameters to obtain a reduction in static torque without causing great increase in dynamic torque.

因此在实践当中通常限制直径比,以减小静转矩又不引起动转矩的明显增大。

2) Employing crawler running mechanism without thrust wheel has reliable property and less maintenance.

采用了无支重轮履带行走机构,性能可靠,维护量小。

3) In the recent models the shearer loaders are self-propelled along a special track without the aid of the chain.

在最新的机型中,采煤滚筒装载机是沿着一条特殊的轨道自推进,而没有链式运输机协助。

4) As such, the bottom of the drive frame is flat, unlike that of the drive head, the bottom of which slopes upward.

正因如此,机尾端传动装置架的底部是平的,不同于那种底部向上斜的机头架。

5) Thus unlike the solid constraint in the frame/chock supports, the pin connections between the legs and the canopy and between the legs and the base

in a shield support make it possible that the angle of inclination of the hydraulic legs varies with the mining heights.

掩护支架不像节式支架或垛式支架那样进行整体固定，它的液压支柱与顶梁之间，以及液压支柱与底座之间，都是销钉连接，这就使得液压支柱的倾斜角度可以随采高的变化而变化。

5. 连词

常用来表达否定意义的连词和连词短语包括 before、unless、until、otherwise、rather than、other、than 等，在英语句中出现，其汉语应译为否定。例如：

1）This development as a pump for special purposes comes from the design of the pump, rather than from any advantage due to size or cost.

这种作为专用泵的趋势主要是由于水泵的结构，而不是由于水泵尺寸和费用方面的优越性。

rather than 和 other than 相当于 not，译成中文“而不”或“不”。

2）Since the flow of water from a centrifugal pump depends on centrifugal force rather than on actual displacement, it can be seen that the output cannot be readily calculated as in the case of the reciprocating pumps.

既然离心泵排出的水流量取决于离心力而与机体实际排量关系不大，显而易见，它的输出量不像活塞泵那样容易计算。

3）It should, however, only be employed in situations where the ground has settled, otherwise it is liable to crack and collapse if used in ground subject to movement.

然而，它（这种拱形顶）只能用于地层已经稳定的地区，如果在不稳定地层使用，便有可能破裂或塌落。

4）It is obviously desirable that the examination required by the order should be made before any shot-hole is charged as, otherwise, gas may be found after charging and the charged shot must then be left unfired.

很明显，按规程要求进行的检查工作，应在炮眼装药以前完成，因为，不这样便有可能在装药后发现瓦斯，已装药的炮眼就只好留下不爆破。

6. 名词

常用来表达否定意义的名词，包括 absence、failure、lack、shortage 等在英语句中出现，其汉语应译为否定。例如：

1）A vacuum, which is the absence of matter, cannot transmit sound.

真空中没有物质，不能传播声音。

absence 本意为“缺乏，缺席”，如果把 absence of matter 译为“缺乏物质”，则不符合汉语表达习惯。

2）In the absence of force，a body will either remain at rest or continue to move with constant speed in a straight line.

如果没有外力的作用，物体或者保持静止，或者继续做匀速直线运动。

3）Hydraulic support operating failure refers to the inaction of some hydraulic support parts in the operation，such as not lifting，moving or advancing.

操作液压支架不动作是指在生产过程中操作液压支架某一部位不动作，如支架不升降、不推溜、不移架等。

4）It might be stated here that a commonly regarded disadvantage of circular shafts，namely，the lack of economy in space for the shaft equipment，might be an advantage where the shaft has to carry a large ventilating current of air.

应当指出的是：大家公认的所谓圆形井筒的缺点，即就井筒装备而言，空间上不经济，也许恰巧是一大优点，因为圆形井筒能通过很大的风量。

7. 短语和句子

1）According to the depth and diameter of different borehole，drilling and blasting method can be divided into shallow hole blasting method，medium-depth hole blasting method and deep-hole blasting method.

根据炮眼深度与直径的不同，钻眼爆破法分为浅孔爆破法、中深孔爆破法和深孔爆破法。

2）For the stable roof，with the exception of the gradually sagging one，its stability must be destroyed artificially and systematically to avoid large areal caving.

对于稳定的顶板来说（不包括逐渐缓慢下沉的顶板），其稳定性还须有计划地进行人工破坏，目的是避免大范围的垮落。

如果将 with the exception of 译为“除了”，则语义不通顺，不符合汉语表达习惯。

3）Centrifugal pump installations are usually located at one point for the life of the mine so greater care should be taken when making the installation.

离心式水泵通常在矿井的整个服务年限中位置不变，所以在进行安装时要更为缜密慎重。

4）The danger in the use of such pumps lies in the tendency of a slip-shod management to use them continuously and thereby create a heavy power loss for the work performed.

滥用这类水泵的危险性在于养成一种粗枝大叶的管理作风：只顾完成任务，不计能量损耗的大小。

5）Congenital fault：a fault caused by the equipment inherent defect in the design or manufacture process.

先天性故障——由于设计、制造不当而造成的设备固有缺陷而引起的故障。

三、英语否定，汉语肯定

否定译作肯定，或称为“反正译”，是指对原文中某些从反面表达的词语，在译文中从正面表达。比如，英文中带有否定前后缀的词语，我们通常把它们译成汉语的肯定形式，例如，disorder 译成“混乱”，carelessness 译成“疏忽”或“粗心大意”，illiterate 译成“文盲”，selfless 译成“忘我”等。与正反译一样，反正译同样可用于对原文中名词、动词、形容词、副词和介词等词语的反译。

1. *动词*

1）It is strictly prohibited that all kinds of waste water directly is discharged into river or lake nearby.

各类废水严禁直接排入矿区附近的河流或湖泊。

2）Cyclic loading and unloading by the support does not affect it.

支架的周期性支撑及卸载对其没有什么影响。

3）Steps of replacing the emulsion pump body：① disintegrate the emulsion pump；② replace new pump body；③ examine and repair the damaged pump.

更换乳化液泵体的步骤：① 解体乳化液泵体；② 更换泵体新品；③ 损坏的泵体升井检修。

4）Scraper conveyor and bridge stage loader should be dismantled in situ and moved to tail part by drawing hoist and are hauled out on flat car. When moving special parts such as head and tail part and transition trough，a special truck is needed.

综采工作面刮板输送机、桥式转载机就地拆除，通过回柱绞车牵引至机尾装平板车外运，特殊件如刮板输送机机头尾推移垫架、过渡槽，需要特殊车辆装车。

5）Operation valve doesn't work or liquid channeling occurs. ① Check whether the operation valve handle is too flexible. ② Check whether there is liquid channeling in the operation valve.

操纵阀失灵或窜液。① 检查操纵阀手柄是否过度灵活。② 检查操纵阀内是否有窜液声响。

6) The fluid coupling contains a limiting fluid port through which a fixed amount of working fluid is introduced to ensure that the output from the fluid coupling will not exceed the maximum value allowable for the motors.

液力联轴节装有一个定量溢流孔，通过该孔使工作液定量流出，以保证液力联轴节的供液量低于电动机所容许的最大流量。

2. 副词

1) Care must be exercised to ensure that the equipment in the shaft does not restrict unduly the flow of air.

要注意保证井筒设备对气流不产生过大的阻力。

2) The ratio of diameters to maintain a constant static torque throughout in wind will become unduly large, particularly in deep shafts.

要想在整个提升过程中维持一个不变的静转矩，那么直径比就会大而失当，特别是对于深井而言。

3) The fault phenomenon is that the cylinder of the balancing jack is off which causes the front beam's elevation is too large and can not draw back.

故障现象：平衡千斤顶脱缸，造成前梁仰角过大，收不回来。

3. 形容词

1) It can be left unsupported for more than 5-8 hr and still remain intact.

这种顶板可在长达5～8个小时内不加支护而保持完整。

2) Another source of unsafe practices is improper motivation and attitude.

违章操作的另一个原因是不良动机和态度。

3) It must be noted that the development trend of each type of hydraulic support has made their application less distinguishable.

必须指出的是各类液压支架的发展趋势使它们在使用方面的区别已变得越来越小。

4) Be sure to have the support, ventilation, lighting and dust prevention to ensure the safe and smooth repair. Ensure the environment is tidy and the passageway unblocked.

保证工作安全、顺利所必需的支护、通风、照明、防尘等措施，并保证环境整洁、过道畅通。

5) The blades may be of special high-strength non-corrosive aluminum alloy.

叶片可以用特殊高强度耐磨铝合金。

4. 名词

1）Entire fault：a fault which makes the equipment or part totally malfunction.

完全性故障——设备或部件完全丧失其应达到的功能的故障。

2）After signed on the discount sheet，the judge should not re-discount on the same item.

裁判员在扣分单上签字后，不得再对该竞赛项目扣分。

3）Misuse failure：a fault caused by wrong assembling，misusing or natural wastage.

使用性故障——由于装配、运行过程使用不当或自然产生的故障。

4）Based on division of labor Judges should be familiar with rules requirement site arrangement and should know well the performance，disassembling，and operation maintenance of various apparatus.

裁判员应根据分工，熟悉规则要求和场地布置，掌握各种仪器装备的性能和使用、拆装及维护保养的方法。

四、双重否定

英语中双重否定句表示强调，一般来说可译作汉语肯定句，但是在有些情况下译作肯定时，语气强度差多了，以 not、without 为例，译为肯定时意思不变，但语气变弱。下面是一些较为典型的双重否定译例。

1）Hardly a month goes by without a word of another survey revealing new depths of scientific illiteracy among US citizens.

美国公民科学知识匮乏的现象日益严重，这种调查报告几乎月月都有。

2）Without careful investigation，you're unlikely come to the right conclusions.

不仔细调查研究，你就会得出错误结论。

3）Without excellent technological skill，the safety would not be guaranteed.

技术不过硬，安全无保证。（技术过硬，安全才有保证。）

4）There is no law that has no exceptions.

凡是规律都有例外。

五、否定的转移

翻译实践中，我们经常会遇到一些否定结构很难直译为汉语，如果按字面意义理解很容易出错，这些结构我们称之为“否定的陷阱”。在矿业英语中，这些否定的陷阱通常为否定的转移。一般情况下，英语中的否定词总是出现在被否定部分的前面，而对其后的部分进行否定。但有时否定词所否定的并不是紧随其后的部分，而是后面的某个部分，这就是否定的转移。例如：

1) In special cases, however, the flexibility of the face conveyor is still not considered sufficiently.

然而在某些场合，工作面输送机的弯曲性能还是显得不够。(not 否定副词 sufficiently)

2) We do not consider melting or boiling to be chemical changes.

我们认为熔化或沸腾不是化学反应。(not 否定的是 to be chemical changes)

3) If we close our eyes, we cannot see anything because our eyelids prevent the rays from entering our eyes.

如果我们闭上眼睛，我们不是因为我们的眼睑阻止光线进入我们的眼睛而看不见任何东西。

not…because/for(并非因为……而)中的 not 属于否定转移，实质上修饰状语从句，可放在 because 之前，从句与从句之间不用逗号隔开。只有在“it is not because…that…”或“not because…do/does/did＋主词＋谓语动词”的句型中，not because 从句才可以放在主句之前。这种句型在翻译成汉语的时候通常译为“并不是因为……而……的”或“并非因为……才……的”。

4) The importance of taking legible field notes and of keeping comprehensive notebooks cannot be overemphasized.

要有清晰的记录，并且保持记录内容全面，其重要性再怎么强调都不过分。

这句话中分别用到了情态动词 cannot 和表示过分的词缀 over，这可以看作“否定之否定”，因此也就成为肯定的意思，即“怎么强调都不可能过分”的意思。

第五节　分译法

英汉两种语言分属两种不同的语系，句子结构存在很多不同之处，因此翻译时往往需要改变句子结构，以适应目的语的表达习惯。分译与合译就是改变原文句子结构的两种翻译技巧。

一、英译汉中的分译法

（一）主语的分译

分译法主要用于英语长句的翻译，为了使译文忠实易懂、通顺流畅，有时不得不把一个长句译成两个或更多的句子。采用这种翻译方法，需要把原文中较长的句子成分，或不易安排的句子成分分出来另作处理，通常译为汉语短句或独立语。例如：

1） Under-expenditure of resources on permanent, high-volume haul roads or over-expenditure on short-term, low-volume haul roads can have serious detrimental effects related to premature failure, compromised health and safety, high maintenance costs for permanent haul roads, or an excessive drain on resources for haul roads with a short, active life span.

无论是对固定的、车流量大的运输道路投入不足，还是对短期的、车流量小的运输道路投入过多，都会造成严重的不利影响，会缩短固定道路的寿命，危害人身健康与安全，增加固定道路维护成本，或是把过多的资源浪费在使用年限短的运输道路上。

该句主语很长，且有连词 or 进行连接，这里分译为由"无论……还……"引导的独立小句。

2） Depending on the size of the operation and the type of haulage system, electric rope shovels, hydraulic excavators, or in some cases large front-end loaders are used in open-pit mining operations.

根据不同矿山规模和运输系统类型，露天开采矿山中会使用钢绳提升电铲、液压挖掘机，有时也会使用大型前端式装载机。

原文句子主语较长，且含有连词 or 引导的短语，所以采用了分译法。

3） These designs allow the mine operator to exchange older components for new, more reliable modules that may have more advanced features.

有了这种设计，矿井运营方可以更换更先进可靠的组件、淘汰掉老化零部件。

该句主语不长，但鉴于谓语动词 allow 的特殊语义，把主语分译为独立小句"有了这种设计"。

4） Examples of terrace mining include Australian coal mines such as Macarthur Coal's Moorvale mine, Peabody's Burton mine, Resources' Jellinbah East mine, and Cockatoo Coal's Baralaba mine.

梯段开采的例子很多，包括许多澳大利亚煤矿，如麦克阿瑟（Macarthur）煤炭的摩尔韦尔(Moorvale)矿山,皮博迪(Peabody)的伯顿(Burton)矿山,杰林巴赫(Jellinbah)资源公司的杰林巴赫东矿,以及考克图(Cockatoo)煤业的巴拿拉巴(Baralaba)矿山。

该句主语为 examples of terrace mining,是复数概念,谓语动词 include 后宾语也较长,这里把主语分译成一个独立小句。

5) If the long axis of the mineralization is not generally north-south or east-west, it may be useful to establish a secondary coordinate system oriented parallel to the long axis.

如果矿化长轴不是南北或者东西方向,那么建立第二个坐标系统,使其平行于长轴,这是非常有用的。

该主从复合句的主句中主语 it 是形式主语,真实主语是后面的动词不定式 to establish…,这里采用分译法译出真实主语,其后通过增加复指代词"这"后跟谓语,译为一独立小句。

(二) 谓语的分译

1. 主语从句的谓语分译

1) It is now generally accepted that coal is of vegetal origin, that the geologic processes which in past ages produced the great seams of coal we mine today are still operating to form deposits which are the basis of coal, and that the several kinds of coal now mined are the result of different degrees of alteration of the original material.

现在,人们已普遍认识到,煤炭是由植物生成的,千万年来的地质过程形成了我们今天可以开采的煤层,这种地质过程今天仍在继续不断地产生着构成煤炭的基础——沉积物,而我们现在开采的煤炭之所以有不同的类别,正是由于原始物料变化程度的不同造成的。

2) It must be emphasised that whilst this ensures greater safety than would be afforded by the use of non-permitted explosives, it does not mean that permitted explosives are incapable of causing ignition of gas or dust in all circumstances that may arise in practice.

必须强调的是:即使这种安全炸药比之于使用非安全炸药能保证较大的安全性,但不等于说安全炸药在实际工作中可能遇到的任何情况下都不会点燃煤尘和瓦斯。

3) With regard to operating techniques, it is generally considered good practice to excavate the farthest or hardest-to-dig material in the first pass while waiting for the truck to position.

关于作业技术方面，在等待卡车就位时，公认最好的做法是：第一斗先挖掘位置最远或最难挖掘的物料。

4）Frequently it is inferred in the mining industry that safety and productivity are inconsistent with one another and that the mining system selected must represent a compromise between the two.

可以猜想，在采矿工业中安全和生产率常是彼此矛盾的，所选择的采煤方法必须把两者统一起来。

5）It is not uncommon that using this method the total methane flow reaches 1,000,000 ft^3/day and the methane emission from the gob is reduced by more than 50%.

常见的情况是，通过采用这种方法，沼气总排放量可达 1 000 000 立方英尺/天，同时，采空区的沼气泄出量可减少 50%以上。

上述例句中带有形式主语 it 的主语从句汉译时，先把主句的谓语译成一个独立语，然后再译从句，或相反为之。

6）What is clear is that, while there is a trend away from the highly developed countries, there has not yet been a major increase in exploration in the highest-risk countries.

很明显，尽管呈现从发达国家转移的趋势，但在高风险国家的勘探仍没有太大增长。

该句含有主语从句 what is clear，其系表结构合成谓语中又含有从句，这里把主语从句独立出来翻译为小句。

2. 被动句中的谓语分译

1）When three or more benches are required, the deposit tends to be treated as massive.

如果一个矿藏需要三个或更多基准，则该矿藏就会当作块状矿床处理。

该句中含有 when 引导的被动句，翻译时谓语分译为带有主语的独立句。

2）The underground mining environment is recognized as being more hazardous than the surface.

人们认为，地下采矿环境比地表更危险。

该句是以动词的动名词短语作主语补足语的被动句，汉译时把谓语分译成一个带有主语的独立句。

3）Furthermore, hollow drill rods are usually employed, down through which compressed air or water is fed to the cutting tool in order to blow out the cuttings.

此外，一般都使用空心钻杆，通过它向下供给压缩空气或水到钻头，以便吹出岩粉。

该被动句主体很短，把谓语分译成一个不带主语的独立句。

4) This is ensured by making certain that all drillings have been removed by the scraper, otherwise, drillings may separate the cartridges and prevent complete detonation of the whole charge.

要使这一措施得到保证，就要确实做到使所有的钻屑用炮眼掏粉勺清除干净，否则钻屑会使药卷之间产生空隙，从而影响到全部炸药的充分起爆。

该被动句主体很短，但由 by 引导的表示施动者的状语很长，因而把谓语分译成一个不带主语的独立句。

5) The importance of taking legible field notes and of keeping comprehensive notebooks cannot be overemphasized.

要有清晰的记录，并且保持记录内容全面，其重要性再强调都不过分。

该被动句主语很长，把谓语分译成一个带有主语的独立句。

6) According to the face conditions encountered, advanced supports were used to reach the outby coal rib as soon as possible.

应根据工作面情况采取超前推进的方式，尽快接近前方煤壁。

该被动句后跟较长的动词不定式作目的状语，这里分译为不带主语的独立句。

3. 后跟多宾语的谓语的分译

1) The advantages and disadvantages of one type of surface mining versus another are often related to the types of equipment used and the associated costs and benefits derived from their use.

一种露天开采法相较于另一种方法的优缺点往往与其采用的设备类型有关，还与设备使用的成本和利润息息相关。

2) Additional low-wall trim will be taken to ensure a stable low-wall angle and a defined low-wall edge offset.

另外还需对底帮进行修整以确保底帮坡脚的稳定，同时保证底帮外缘明确清晰。

3) The stacked configuration can result in low rehandle (25% or less) and high productivity, especially for narrow strips.

采场覆盖式开采法可以减低重复剥采量(25%或更少)，提高生产力，尤其是对狭长的采掘带而言。

4) Nevertheless, the extraction of natural resources now attracts the sort

of responsibility and scrutiny that few other global industries are subjected to.

然而，开采自然资源如今同全球其他行业相比，会带来更大责任，要经受更严格的审查。

上述四例中，谓语动词一词多义且后跟有多个宾语，可采取分译法，根据搭配意义分开表述。

（三）状语的分译

1）Petroleum was formed at a later time than coal from plants and animals that lived in shallow seas.

石油是由生长在浅海的动植物形成的，其形成时间比煤晚。

2）Sequentially it involves excavating a block of overburden 30 m long to create the new highwall (the key) and using this overburden to build a working bench out into the previous strip's void.

整个顺序依次如下：开凿一个 30 米长的剥离物区域，露出新的顶帮（切槽），使用这部分剥离物建造一个延伸至前一个采空区的工作台阶。

3）Briefly, the longwall system of working consists of extracting the whole of the coal in a single operation from a more or less continuous and long working face or from a number of suitably disposed long faces, hence its name.

简言之，长壁采煤法是从一个比较连续的长壁工作面，或从布局合适的几个长壁工作面，一次作业采完全部煤炭，这就是本名称的由来。

4）Bucket fill factors of more than 100% are achieved by heaping material in the bucket.

铲斗内可以堆装物料，因此满斗系数可以超过 100%。

5）For health and safety reasons, safety berms (also known as safety benches or windows) are constructed along crests of benches in a similar manner to those found next to haul roads.

出于对健康与安全因素的考虑，此处的安全平台（也可被称为安全台阶或挡墙）是沿着台阶的坡顶线建造的，这种方式与运输道路旁安全平台的建造方式类似。

6）The space from which the coal is removed, except for roadways, may be packed either wholly or partially, depending largely on the amount of debris available or to be disposed of.

采煤后，所留下的空间，除巷道外，可全部或部分充填，这在很大程度上取决于采后留下或需要处理的矸石量。

7) The discovery that certain black rock would burn was undoubtedly accidental and probably occurred independently and many times in the world over thousands of years.

发现某种黑色岩石能够燃烧毫无疑问是偶然的，而且在过去几千年中多地、多次有此发现。

8) The longer the useful life, the more obsolete the equipment becomes in comparison to new technological advances.

设备的使用年限越长，就会越老化，而科技发展却日新月异。

上述八例中英语原文的状语太长或作状语的副词短语不宜译为相应的汉语状语，因而将它们分译为小句。

9) The whole area was sinking slowly, at such a rate that for most of the time the sand, mud and clay deposited by the rivers on the bottom of the estuary kept the depth of the water fairly constant.

整个地带在缓慢地下沉，其下沉速度恰好使江河冲来沉积在河口底部的泥、砂和黏土经常保持水的深度不变。

该句状语 at such a rate that 非常之长，这里通过添加主语“其”的方式将它译为带有主语的独立句，清晰明了。

10) Between 1980 and 1992, 136 abandoned coal mine sites in Pennsylvania were reclaimed, at a cost of about $2 348 per hectare.

1980 年至 1992 年期间，宾夕法尼亚州的 136 个废弃煤矿场被开垦，每公顷耗资约 2 348 美元。

该句状语 at a cost of 较长，这里译为主、谓、宾完整的独立句，清晰流畅。

二、汉译英中的分译法

这里又要提到两种语言的差异。汉语句子以“意合”为特点，由于它没有词形变化(动词尤其突出)、定语从句和独立主格，连接词和介词也比英语少得多，所以往往用隐性连贯方法来表示各种语法关系，句子形式比较松散、自由。英语句子重“形合”，往往用词形变化，连接词、介词、定语从句和独立主格表示成分之间各种语法关系，句子形式很严谨。因此在汉英句型转换时可以采用词形变化、译成并列句的形式等方法进行分译。

(一) 按内容层次分译

1) 应力对红外辐射具有很好的控制作用，且潮湿砂岩的控制效应比干燥砂岩更强，为采用 VDIIT 指标反映砂岩单轴加载过程中的应力变化情况奠定了基础。

Accordingly, it can be concluded that the application of stress would impose favorable control effect on IR, especially for wet sandstones. This can lay the foundation of reflecting stress change using VDIIT index during uniaxial loading process.

2）放顶煤开采具有巷道掘进率低、产量大、效率高、开采成本低等优点，近20年来在我国得到迅速发展，并且已经成为厚煤层开采的主要方法之一，为煤炭企业带来了巨大经济效益。

Since LTCC has the advantages of high-production, high-efficiency, low development ratio, and low-cost, it has developed rapidly in the past two decades and become the most extensively used mining method. It has brought huge economic benefits to coal industry.

以上两句有着相似的结构特点，均用“为”字引出第二层意思，表示第一层意思的结果，因此分别译为由 this 和 it 引出的第二句。

3）根据济三煤矿试验区域采矿地质条件及南阳湖堤的结构特点，确定了湖堤的变形极值，分析得出控制湖堤变形的关键因素为工作面充实率和等价采高。

Based on the mining and geological conditions of Ji San Coal Mine test site and the special features of Nan Yang Lake dam, the limiting deformation values of the dam were determined. Analysis of the data showed that the key factors controlling the deformation of the dam were the backfilling ratio and equivalent mining height at the face.

该句有两层意思，因此译文在第二层内容处断开，译为含有不同主语的两个被动句。

4）托盘是锚杆的重要组成部分，对锚杆作用的发挥有重要影响。

The bearing plate is one of the most important parts of the bolt. It plays an important role in developing the supporting function of the bolt.

该句有两层意思，因此译文在第二层内容处断开，译为含有名词主语和指示代词主语的两个句子。

5）大倾角大采高开采煤层后应力重新分配，在工作面煤壁前方的一定范围内形成超前支承压力影响区，支承压力超过煤体塑性极限时煤体将产生裂隙、破断，支承压力向煤壁深部转移，在煤体承载力与超前支承压力达到极限平衡时，煤体才处于稳定状态。

Stress is redistributed after the steeply dipping coal seam of large height is mined. The abutment pressure influence area is formed within a certain distance in front of the coal face. In this area, the coal seam is cracked and broken when the abutment pressure exceeds the coal plastic limit. The abutment pressure moves far into the panel coal body, and the coal body becomes stable when the load capacity of coal and the front abutment pressure reaches the state of limiting equilibrium.

该句有四层意思，因此译文在第二、三、四层内容处断开，译为四个句子。其中第二层意思译为被动句；第四层意思表述较长且含有较长的时间状语，译为when 引导的时间状语从句的并列句。

6）当观测点位于工作面前方一段距离时，开始出现下沉并随着工作面的推进继续发展：工作面推过观测点后，其下沉先加速后减速，直到观测点位于工作面后方一段距离时下沉停止。

When the observation point is some distance ahead of the face, it begins and continues to subside as the face continues to advance. After the face passes the observation point, subsidence accelerates, followed by deceleration. Subsidence stops when the observation point is located some distance behind the face.

该句第二层意思由冒号引出，而冒号后又含有两层意思，因而英译为三个完整的句子，每个句子又基于原文内容和目的语的结构特点进行表述。

（二）从主语变换处分译

1）红外辐射观测设备采用 FLIR A615 红外热像仪，热灵敏度（NETD）＜0.05 摄氏度，分辨率为 640×480 像素，像素间距 17 微米，图像采集速率为25 帧/秒，波长范围为 7.5～14 微米。

FLIR A615 IR imager was used for IR observation and the specific testing parameters are listed below. The thermal sensitivity was below 0.05 ℃, the resolution was 640×480 pixel, the pixel pitch was 17 μm, image acquisition rate was set as 25 frames/s, and the wave-length range was 7.5-14 μm.

该句第一个逗号后的主语发生变换，具体为“FLIR A615 红外热像仪”的“热灵敏度”，故而在此采用分译法，逗号前后各译为一句。

2）UDEC 软件中的裂隙可以直观表示，图 7 中的红色表示煤层开采后的覆岩裂隙。

The distribution of fractures in the model was visualized by UDEC. Mining-induced fractures in the overburden are shown in the red region in Fig. 7.

3）从 20 世纪 70 年代开始，综合机械化采煤的应用和推广大大加快了工作面的推进速度，使产量大幅度增长，瓦斯涌出量也大大增加，瓦斯超限成了全国各瓦斯矿井普遍面临的问题。

Since the 1970s, the application and wide use of longwall mining has greatly increased the panel advancing speed, coal production, and methane emission as well. Therefore, excessive gas emission becomes the common problem in all gassy mines.

以上两句均有两个主语，翻译时均在第二个主语处另起一句。

4）喷浆从工作面后方 40 米处开始，喷浆材料为水泥∶沙子∶石子＝1∶2∶2；速凝剂为水泥重量的 5％～8％。

Grouting begins 40 m inby the face. The grout materials consist of cement∶sand∶gravel＝1∶2∶2. The accelerator used is 5％-8％ of cement weight.

该句含有三个主语，翻译为三个不同主语的英文句子。

（三）从关联词（如转折）处分译

1）但大量工程实践表明，许多情况下现有巷道围岩控制技术仍不能有效地控制高应力软岩巷道的强烈变形，而且造成这种情况的原因不是支护强度不够，而是由于现有支护承载结构的稳定性较差，难以有效控制巷道围岩的强烈变形。

However, a large number of engineering practices show that the existing rock control technologies are unable to control the large deformation of the soft rock roadways effectively in many cases due to high stress. The reason for the large deformation of soft rock roadways is usually due to the poor structure stability of the existing frame bearing structure that can not effectively control the severe deformation of entry surrounding rock rather than insufficient supporting strengths.

该句在关联词“而且”处采用分译法，后面的原因部分另起一句。

2）地下开采的产量比例占 90％，即使是未来露天开采有所发展，地下开采的比例也不会低于 85％，而且地下开采基本都是长壁开采。

Underground coal mining produced about 90％ of total production. Even if in the future open-pit coal mining becomes fully developed, the percentage of underground coal mining production will not be less than 85％ of the total. Longwall mining is the most widely used method for underground coal mining.

该句在关联词“而且”处采用分译法，后面的强调部分另起一句。

3）EH-4 电导率成像系统以其数据采集量大、设备轻、速度快、精度高、工作效率高、探测成本较低等优点，在探水、探采空区中得到了广泛的应用，但 EH-4 在复杂巨厚煤层条件下，导水裂隙带及离层空间等覆岩破坏范围确定方面的应用较少。

The EH-4 electrical conductivity imaging system has gained wide application for water and coal mine gob exploration due to its significant advantages, such as large data collection capacity, light weight, fast analysis, high precision, high efficiency and low cost. However, the EH-4 system is rarely used for defining the overburden failure range, including the development of water flowing fractured zone and the bed separation space under the complex conditions of extra-thick coal seam.

该句在关联词"但"处采用分译法，后面的转折部分另起一句。

4) 针对近距煤层群采动覆岩裂隙的发育规律，张玉军和胡永忠等发现下层煤的重复开采会造成新的裂隙产生和原有裂隙再次发育，导致重复开采扰动区导水裂隙发育高度增加，且裂隙发育高度受到煤层的复合厚度、层间距和开采顺序等因素制约。

Zhang Yujun and Hu Yongzhong found that the HCF zone's height in the overburden increased when continuously mining the lower seam due to the generation of new fractures and reactivation of existing fractures. Moreover, multiple factors, including the comprehensive thickness of the coal seams, distance between adjacent seams, and mining sequence, all affected the height of the HCF zone.

该句在关联词"且"处采用分译法，后面的强调部分另起一句。

5) 砂岩试样在单轴加载过程中，其 VDIIT 指标虽可用于预测判断岩石的破裂或失稳前兆，但事实上 VDIIT 突变并非与 SR 突变完全同步，而是存在一定的滞后性。

During the uniaxial loading process, VDIIT can serve as the precursor of rock failure or instability. It is worth noting that VDIIT mutation is not in complete synchronization with SR mutation, but lags behind it.

该句在关联词"但"处采用分译法，后面的转折部分另起一句。

6) 在大量现场实测与理论研究的基础上，建立了下分层综放开采的覆岩结构模型，并对大、小结构的稳定性及其对矿压显现的影响进行了分析。

Based on lots of field measurements and theoretical analysis, the model of overburden structure for the fully mechanized multi-slice top coal caving was established. It has been used to analyze the stability of overburden structures large or small and their effect on mine pressure.

该句在关联词"并"处采用分译法，后面的并列部分另起一句。

（四）从意义完整、独立处分译

1）地表下沉预计可以用来评价可能的下沉量对地面建筑物和环境的影响，也可以用来改进煤矿设计、采煤方法或者开采顺序，以此来降低下沉影响的严重性，便于设计和采取减沉措施。

Surface subsidence prediction can be used to evaluate the influence of possible subsidence on surface structures and the environment, and can also be used to improve the design of a coal mine, mining methods and mining sequence. Therefore, it can reduce the severity of subsidence influence, and facilitate the design and selection of measures for reduction of surface subsidence.

该句在“以此”前后意义完整、独立，因此在这里采用分译法。

2）近年来，作者研究了浅埋近距煤层采动覆岩导水裂隙发育规律及机理，发现了不同开采区域的覆岩导水裂隙发育特征，提出了工作面错位布置、降低采高、加快工作面推进速度和局部充填等控制重复开采扰动区导水裂隙高度的方法。

Recent studies have revealed the pattern and development mechanism of HCF in overburden induced by mining shallowly buried coal seams that are near each other. In addition, these studies provided several approaches to control the HCF zone's height, such as arranging the faces in a staggered pattern, reducing the mining height, increasing the advance rate of the face, and using partial backfill mining.

该句在“提出了”前后意义完整、独立，因此在这里采用分译法。

3）同时，微山湖地处中国南水北调的重点生态保护区，如何在安全采出湖下煤炭资源的同时，最大限度减小开采对湖区水资源的影响，即实现保水采煤，是该矿区亟待解决的难题。

It is a key nature reserve along the eastern route of the China's South-to-north water diversion project. It is urgent in this area to minimize the impact of mining on local water resources while ensuring safety, namely to achieve WCM.

该句在“如何”前后意义完整、独立，因此在这里采用分译法。

4）岩石受力破坏过程中红外辐射会发生变化，对岩石红外辐射进行监测，可以分析岩石的变形和破坏特征，并有可能为预测岩石的破坏前兆提供可靠信息。

The infrared radiation（IR）on the surface of rock will change during loading process. By monitoring the IR, the deformation and failure characteristics of rock can be obtained, and reliable information for predicting rock failure precursors may be obtained, as well.

该句在第一个分句后意义完整、独立，因此在这里采用分译法。

（五）原文出现总说或分述时要分译

1）实验选取四种煤样，分别为乌兰煤样，补连塔煤样，口孜东煤样和大佛寺煤样。乌兰和补连塔煤样（标记为 WL 和 BLT），分别来自内蒙古乌兰哈达煤矿和内蒙古补连塔煤矿。口孜东煤样（标记为 KZD），来自安徽口孜东煤矿。大佛寺煤样（标记为 DFS），采自陕西大佛寺煤矿。

Four coal samples were selected for the experiment from Ulanhada Coal Mine and Bulianta Coal Mine in Inner Mongolia Autonomous Region, East Kouzi Coal Mine in Anhui Province, and Dafosi Mine in Shaanxi Province. They were labeled as WL, BLT, KZD and DFS, respectively.

该句在短小的总说后进行分述，译文采用分译法，把总说和分述巧妙糅合，译文简洁流畅。

2）支撑掩护式支架的顶梁长度一般均较掩护式支架长，增加了支撑力，缩短了掩护梁长度，加大了掩护梁的坡度，同时还改善了支架底座与底板受力情况。

Compared to the shield support, the chock-shield support has a longer canopy with increasing leg loading capacity but decreasing length and increasing inclination of the caving shield. Generally, the load distribution of the support base and floor is improved.

该句原文在“同时”处含有概述之意，因而使用 generally 分译为独立一句。

（六）为了强调语气而采用分译

1）充填开采技术已经基本成熟，在中国许多矿区都得到了应用，目前存在的主要问题是产量低、成本高，尤其是在目前煤炭经济形势不好的情况下，还不能进行大规模的推广应用。

Coal mine backfill, as a validated technique, has been increasingly used in many mining areas in China. However, its existing problems are low productivity, and the recent depressed coal market has limited its large-scale application.

2）当上、下煤层距离较近时，上、下煤层开采形成的垮落带和裂隙带范围可能重叠一部分，重叠的范围取决于上、下煤层的层间距。

When the upper and lower coal seams are near each other, the corresponding caved and fractured zones induced by mining the upper and lower seams may partially overlap. The scope of overlap depends on the distance between the two seams.

3）由于中国的煤炭开采量巨大，条件又相对复杂，也导致中国的煤炭开采技术多样化，对于不同的煤层条件具有不同的开采技术和开采特点。

Due to the large amount of coal production in China from various complex mining conditions, mining methods vary considerably. Consequently there are different mining methods with different special features for different coal reserves.

以上三句的特点是后一部分有明显强调的意味，所以后一部分分开来译，以更好地传达原文的语气。

第六节　合译法

一、英译汉中的合译法

合译法主要用于定语从句，即把定语从句与主句合译成一句。英译汉时，把定语从句译成句中的一个成分，主要是译成定语或谓语。

（一）译成定语

英语中的限制性定语从句，均位于它所修饰的词之后，一般为全句必不可少的成分，如果省去，主句就会发生意义上的变化，甚至由于意义不完全而不知所云。所以，凡是结构不太长的限制性定语从句，常常把从句融合在主句里，省去关联词，译成定语，即"……的"，因为限制性定语从句与所修饰的词关系密切，分开译往往影响主句意思的完整。某些结构短的非限制性定语从句，有时也可译为定语。例如：

1）The first theory, known as the in situ or "in place" theory, proposes that peat beds were formed from deposits which accumulated through the vegetal matter falling at the place where it grew.

第一种理论，即所谓就地生成论，认为泥煤层是由在生长地倒下的植物体堆积而成。

2）If the seam being mined has a high methane content, the mining method employed should be those that extract with high recovery and leaves as little coal in the gob as possible.

如果被开采的煤层具有很高的沼气含量，就应该采用那些回收率高，尽可能不在采空区留下煤的开采方法。

3) Therefore, the unloading end of a stage loader where it dumps coal into the entry belt conveyor is raised and can be pushed to overlap the entry belt conveyor.

因此,将煤卸载到平巷带式输送机的转载机卸载端可以升起,并能向前推移,与平巷带式输送机搭接。

4) On this shift the coal is cut, the props and bars that supported the roof of the previous day are removed and the conveyor in the main gate is extended so that the coal from the face conveyors which have been moved, will still pour onto it.

该班任务为采煤,拆除前一天支护顶板的支柱和顶梁,并接长平巷输送机,以便使已向前移置的工作面输送机运出的煤仍将持续不断地卸载到平巷输送机里。

5) This seed is stored until the wet season when it is broadcast over areas that have been backfilled and covered with topsoil.

这些种子储存至雨季,再播种于已回填且覆盖了表土的区域。

6) To forget the contribution that mining has made (and continues to make) is to take for granted the significant progress that civilization has made since the last Ice Age.

忘记采矿已作的贡献(而且在继续作贡献),就是把上个冰河时代以来文明所取得的重大进步视为理所当然。

7) As the seams progress to greater depths, typically at around 85 m, a truck and loader pre-stripped operation must be introduced to assist the dragline to continue to reach the top of the coal by creating a working level for the dragline that is lower than the natural topography.

随着煤层采掘深度的增加,通常在 85 米左右,就必须引进卡车和装载机进行预先剥离台阶辅助作业,通过创建一个低于自然地平面的拉斗铲工作平面,来协助拉斗铲继续作业以达到煤层顶板。

8) Although it may be too much to expect the world at large to respect mining engineering as a profession, its citizens may in time acknowledge the essential role that mining performs and the constraints under which it operates.

尽管全世界普遍把采矿工程尊为一种职业这种期望可能过高,但其公民迟早会认识到采矿所发挥的根本性作用及其所受的多种限制。

9) This is another proof that the speed of sound depends in general on the

temperature of the substance through which it is passing.

这再次证明，声音的速率一般取决于它所经过的物质的温度。

10) In as much as most classifications of coal throughout the world are based on similar criteria, it is generally practical to compare coals described in systems used elsewhere with coals whose rank is described in the American system.

由于全世界大多数煤的分类都基于类似的标准，所以将其他地方所采用的分类法描述的煤等级与用美国分类法描述的煤等级进行比较总是可行的。

11) They are defined as dislocations in which the hanging wall is displace downwards relative to the foot-wall.

它们被定义为上盘相对于下盘向下移动的错位关系。

12) In longwall coal mining, the coal cut down by the shearer or the plow is transported by the armored face chain conveyor (AFC), which is laid across the full face width, to the headentry T-junction, where it is transferred to the entry belt conveyor through the mobile stage loader.

在长壁工作面采煤时，滚筒采煤机或刨煤机采下的煤，通过贯穿工作面全长铺设的铠装刮板输送机(AFC)，运送到工作面运输巷 T 字形连接处，然后再由移动式转载机转载到运输平巷的胶带输送机上。

上述十二例中，前十一例均把限制性定语从句译成"的"字结构；而例 12)则把原文中的非限制性定语从句译成了"的"字结构。

(二) 译成谓语

1) The economic seams are contained in the Mid-to-Late Permian German Creek Formation, which are overlain by up to 55 m of poorly consolidated cemented sediments consisting predominantly of sand and clay with irregular gravel beds and weathered basalt flows.

经济可采煤层存在于中晚期二叠纪德国克里克地层，上覆 55 米厚的不完全固结胶结沉积层，主要由砂石和黏土构成，同时伴有不规则的砂砾和风化玄武岩。

该句译文很有特色，因为句子的重点是在从句上，为了突出定语从句的内容，所以把定语从句译成谓语，而主句压缩成汉语词组作主语。

2) It has the disadvantages that there are limits to production because it is cyclic mining, that is, it involves separate operations which includes cutting the coal, drilling the shot holes, charging and shooting the holes, loading the broken coal, and installing roof supports, as enumerated above.

普采法的缺点是由于循环开采，所以产量受到限制，它包括多项独立进行的作业，如以上列举的截煤、钻眼、装药、爆破、装运爆破下来的煤，以及进行顶板支护等。

本句主要采用合译。首先主句中 that 引导的定语从句译成了谓语，其次由于原句是由 that is 连接的语义关联密切的两句，译文中翻译成逗号相连的一句。

3) Moreover, factories, chemical and food processing plants, and thermal power plants, which are generally closer to population centers than are mines, sometimes produce more pollution than do mines.

此外，工厂、化工厂和食品加工厂以及火力发电厂通常比矿山离人口密集区更近，有时产生的污染比矿山更严重。

该句采用合译法，把 which 引导的非限制性定语从句译成谓语。

4) Coal is a rock derived from wood and other plant tissues that flourished several hundred million years ago.

煤是数亿年前旺盛生长的树木和其他植物组织变成的一种岩石。

5) In order to minimize the recirculation of coal fines which can be a serious problem with the AFC, there should be an adequate height difference at the transfer point between AFC and stage loader.

粉煤的回流会成为工作面铠装链板输送机的严重问题，为使粉煤回流减少到最低限度，转载点处的工作面铠装链板输送机与转载机之间应具有足够的高差。

以上两句中限制性定语从句均译为谓语。

6) Each section is constructed from two special side rollings that are welded to a deck plate, which forms the surface for the conveying chain assemblies.

每个溜槽段由两个特制的滚轧槽帮焊接在溜槽中隔板上而成，这样溜槽中隔板就形成了链板输送装置的层面。

该句中 that 引导的限制性定语从句使用合译法，译为谓语；由 which 引导的非限制性定语从句则使用分译法译出。

7) Therefore, they can still transmit horizontal forces, although the rear end of the strata, which is in the gob area, is generally lower than the front end, which is located above the powered support and the immediate roof.

因此，虽然位于采空区的岩层后端一般低于液压支架及直接顶上方的岩层前端，它们（这些岩石）依然能够传递水平推力。

该句中含有两个 which 引导的非限制性定语从句，这里译为谓语，简洁明了。

8) For single coal seams that occur deeper than 30 m and where economies of scale allow, a single-seam dragline method is often applied.

对于深度超过 30 米的单层煤层，在经济规模允许的情况下，通常采用单煤层拉斗铲作业法。

该句中 that 引导的限制性定语从句使用合译法，译为谓语。

二、汉译英中的合译法

英语长句使用较广泛，从句套从句、短语含短语的情况较多，而汉语则较多地采用小短句，所以在汉译英时，就要把汉语的两个句子甚至更多句子合译为英语的一句。具体方法如下。

(一) 在复指代词处合译

1) 这两个工作面倾角为 0°～5°，其对覆岩导水裂隙的发育影响较小。

The inclined angle of the two working faces ranges from 0° to 5°, which has little effect on HCF development.

原句中的分句主语为指示代词“其”，这里译成 which 引导的非限制性定语从句，从而合译成主从复合句。

2) 其中，EH-4 大地电磁法常用于地球电性和磁场探究，它利用天然电磁场作为场源。

For example, the EH-4 magnetotelluric method that uses the natural electro-magnetic field as a magnetic field source is often used to explore the electric and magnetic fields on the earth.

原句中的分句主语为指示代词“它”，这里译成 that 引导的限制性定语从句，从而合译成主从复合句。

3) 这种影响可归因于这样一个事实，煤样尺寸越大，包含的缺陷就越多，因此煤样的强度就越低。

This effect is commonly attributed to the fact that the larger the specimen the more defects it contains, and thus the lower the strength.

该句中含有复指代词“这样”，这里采用合译，译为名词加上同位语从句的形式。

4) 1775 年，詹姆斯·瓦特在英国发明了第一台蒸汽机用来抽煤矿中的水，这是一个非常重要的应用，使得矿井有可能开采更深的部分。

The first steam engine was invented by James Watt in 1775 in Britain to pump water from coal mines, a very important application that made it possible for mines to go deeper.

该句中含有复指代词"这",这里采用合译,译为名词同位语加上定语从句的形式。

(二) 按内容连贯合译

1) 许多研究已证明随着工作面的推进,顶板对液压支架将产生很大的压力。尤其在顶板周期来压和初次来压过程中,工作面的液压支架增阻较大。

Many researchers have demonstrated that during the advancement of the longwall face, the roof applies a large pressure on the shields, and the roof pressure is greatly increased during first roof weighting and periodic roof weighting.

该句中使用"尤其",前后两句内容连贯,合译为 and 连接的并列句。

2) 制样时,首先将煤样进行取芯处理,然后破碎、研磨、筛分成不同的粒径备用。

During sample preparation, cores of coal were taken and then smashed, ground and screened into particles with different sizes for later use.

原句意思连贯,一个主语后跟多个谓语,译成英语的并列谓语。

3) 实验采用的砂岩试样取自山东某煤矿,且采用同一整块岩石加工而成。试样加工尺寸为 70×70×140 毫米。

The sandstone samples with an identical size of 70 × 70 × 140 mm were obtained from a same rock of a Shandong coal mine.

该句虽有两个句号,但均为对主语的描述,内容连贯,合译为一句。

4) 井筒泛指立井和斜井,也包括暗井。通常由井颈、井身和井窝组成。

Shaft mainly includes vertical shaft, inclined shaft and blind shaft, and it is usually composed of shaft collar, shaft body and shaft sump.

该句虽有两个句号,但是是对同一主语的描述,属并列关系,合译为一句。

(三) 从主语变换处合译,常译为主从句

1) 整个反应是一个连续的循环反应过程,两种反应的综合作用导致煤体热量的快速集聚和不可控自燃的发生。

The whole reaction is a continuous and cyclic process where combined effects of the two reactions cause the rapid accumulation of heat and trigger uncontrollable coal spontaneous combustion.

该句中出现主语变换,在变换处译成 where 引导的定语从句。

2) 实验加载设备采用 C64.106/1 000 千牛型电液伺服万能试验机,最大载荷 1 000 千牛。

The electro-hydraulic servo universal tester (C64.106/1 000 kN), with a maximum load of 1 000 kN, was used in the loading process.

该句在主语变换处采用合译法，译成 with 短语作定语。

3) 该矿年产量 3 公吨，开采 31 煤和 32 煤，31 煤层平均厚 5.0 米，32 煤层平均厚 4.0 米。

It has an annual production of 3.0 Mt, mining coal seams #31 and #32, which have average thicknesses of 5.0 and 4.0 m, respectively.

该句翻译较为灵活，根据其各个部分的关系进行组合，根据语义关系译成 which 引导的非限制性定语从句。

许多翻译教程中不把分译与合译列为独立的翻译技巧，但它们常见于某些句子成分、短语和从句的翻译中。

第七节　从句的翻译

在矿业英语中，简单句的使用相对比较少，更多使用复杂长句来表述客观、准确、严谨的科学概念，因此在翻译复杂长句时，首先应理解从句之间的关系和翻译时它们所处的位置，才能使译文准确、通顺、流畅，符合汉语的行文习惯。按照从句在句子中的作用分类，从句可以分为主语从句、宾语从句、表语从句、定语从句、同位语从句、状语从句，其中主语从句、宾语从句、表语从句和同位语从句在整个句子中起名词作用，故统称为名词性从句。本节将对这几种相对独立的从句类型的翻译方法逐一进行介绍。

一、名词性从句的译法

英语名词性从句包括主语从句、宾语从句、表语从句和同位语从句四大类，这些句子或是在句子中作主语，或是宾语、表语、同位语，在句子中的作用相当于名词，因此常被称为名词性从句。这类从句的翻译比较灵活，大多数可按原文的句子顺序译成对应的汉语，但在有些情况下，也可视上下文使用一些其他的处理方法。

（一）主语从句的翻译

主语从句是指在主句中充当主语成分的句子。主语从句主要有两种形式：一种是从句位于主句主语的位置上，即“主语从句＋谓语＋其他成分”；另一种是以 it 为先行词作形式主语，真正的主语从句在主句的谓语之后，即“it＋谓语＋主语从句”。

1. 主语从句位于句首

主语从句放在句首，由关联词或从属连词直接引入主语，翻译时一般可按原文顺序翻译。主从复合句的从句的引导词常有 what、which、how、why、where、who、whatever、whoever、whenever、wherever 以及从属连词 that、whether、if。例如：

1）That coal has formed from accumulations of plant material has been well established.

煤是由植物类物质的堆积形成的，这已经得到了充分的证实。

2）Whether an unstable slope results in significant cost to the operation depends on the rate of movement, the type of mining operation, and the relationship of unstable material to the mining operation.

不稳固边坡是否会使矿山承担很大的花费，取决于位移的速度，采矿作业类型以及不稳固岩石与采矿作业的关系。

2. 以 it 作形式主语的主语从句

以 it 作形式主语所引出的真正的主语从句，可以采取顺译法和倒译法进行翻译。

所谓顺译法，就是按照原文中句子的词序翻译，即先译主句，通常译成无人称句，it 不需要译出来，或增译"人们""我们""有人"等词语，然后再译主语从句。

相反，倒译法就是把真正的主语从句先译出，为了强调，it 可以译出来，通常译作"这……"，如果不需要强调，it 也可以不译出来。无论采取顺译还是倒译的翻译方法，目的都是使译文更加流畅，重点突出，符合汉语的行文规则。例如：

1）It will be obvious that the primary precaution against ignition of firedamp is adequate ventilation.

很明显，预防引爆瓦斯的主要措施是保证良好的通风。

2）It is strictly prohibited that all kinds of waste water is directly discharged into the river or lake near the mining areas.

严禁各类废水直接排入矿区附近的河流或湖泊。

3）There is no doubt that many early and disastrous explosions were due almost entirely to firedamp and arose from defective and inadequate ventilation.

毋庸置疑，许多早期的灾难性爆炸几乎都是由于瓦斯及通风系统的缺陷，供风不足导致的。

4）It is being increasingly recognized that life and health of the worker are the greatest assets of industry, far more valuable than any other factor.

人们越来越深刻地认识到，劳动者的生命和健康才是产业中最大的财富，其价值远远超过任何其他方面。

5) Methods for converting coal to liquid and gaseous fuels are known, and it seems certain that such manufactured fuels will become important as supplements to petroleum and to natural gas.

把煤炭变成液体燃料和气体燃料的方法已研究出来，因此作为石油和天然气的代用品，加工后的煤燃料会具有重要意义，这一点是必然的。

以上例句，例 1)到例 4)采用的是顺译的翻译方法，前三句表达的是事实、常理或规定，故采用了译成无人称句的方法，例 4)表达了人们所认识到的事实，为了使译文成分完整，采取了增译出人称主语的方法。例 5)是个复合句，第二小句中的实际主语是 that 引导的主语从句，it 是它的形式主语，为了使译文意思表达更加流畅，采取了倒译法，把 it seems certain that 译为“这一点是必然的”，放在了句尾。

(二) 宾语从句的翻译

在主句中作宾语成分的句子叫作宾语从句。英语的宾语从句有两种：一是动词的宾语从句，包括不定式、分词、动名词后的宾语从句；二是介词后的宾语从句。翻译宾语从句时，无论是动词宾语从句还是介词宾语从句大都采用顺译法，即按照原文的顺序。

1. 动词引导的宾语从句

1) European governments require that all possible deposits be mined to conserve the nation's energy resources.

欧洲各国政府要求采出所有可采的煤层以保护国家能源。

宾语从句中，有些表示意愿的动词要求后面的宾语从句必须使用虚拟语气，本句中 require 就是这种用法，其他表示意愿的动词还有 order、command、desire、demand、request、promise、insist、arrange、recommend、propose、suggest、prescribe 等。

2) Careful study of mining operation shows that safety and efficiency are intimately related.

对矿业的深入研究表明，安全和效率密切相关。

3) There is no generic way to determine whether an operation would benefit economically from a trolley-assist system.

没有通用的方法可以确定一项作业是否能从架线式辅助运输系统中获得经济利益。

4) It is well to remember that, on a longwall face, all the powered

supports function as one complex multi-unit machine to form a single strata control system.

最好要记住，一个长壁工作面中的全部液压支架是作为一个多单元的复合系统而起作用的，这样就形成了一个整体的地层控制系统。

5) The supervisor should understand why men act in an unsafe manner and how unsafe conditions are produced, so he will be able to recognize and correct both situations before they lead to an accident.

管理人员应该懂得劳工为什么以不安全的方法操作和不安全条件是如何产生的，因此，管理人员在这两种情况引发事故前应能够认识它们并予以纠正。

本句中动词 understand 的宾语成分是两个宾语从句，分别由 why 和 how 引导。

以上例句的宾语从句都是出现在动词后，翻译时大都按照原文顺序翻译即可。

2. 介词引导的宾语从句

介词宾语从句前面的介词一般和动词、形容词或副词有关。翻译时，顺序一般不变。例如：

1) The decision as to whether a slope or shaft will be employed as a portal can be a very difficult one.

决定是用斜井还是立井做井筒是很难的。

本句中介词短语 as to 后的宾语成分是由 whether 引导的句子，采用了顺译的翻译方法。

2) The ultimate disposal of these treated waste waters is a function of whatever local, state and federal regulations may apply in any situation.

这些处理过的废水最终如何处理，取决于在任何情况下都适用的地方的、州的或联邦的立法条例。

3) There are examples of where there is no immediate roof as the massive main roof lies directly on the coal bed, but this is rare.

也存在由于厚层老顶直接位于煤层之上而不存在直接顶的例子，但这毕竟很少见。

句中 of where there is no immediate roof as the massive main roof lies directly on the coal bed 为介词短语，修饰 examples；在这个介词短语中 where there is no immediate roof as the massive main roof lies directly on the coal bed 为介词引导的宾语从句；在这个宾语从句中，又有用 as 引导的原因状语从句 the massive main roof lies directly on the coal bed。

4）As seen there are two groups of thyristors which are fired according to whether the load voltage is to be positive or negative.

可见有两组半导体开关元件，他们分别是根据负载电压是正还是负而受到触发。

本句中 according to 是介词，后面的 whether the load voltage is to be positive or negative 是介词引导的宾语从句。

位于介词 except、but、besides 后的宾语从句，汉译时通常要前置，可以译为一个并列分句，用“除了”“除……之外”“只是”“但”等表示，但有些时候为了行文的流畅自然，宾语从句也可以后置，要视具体情况而定。例如：

5）The lift component is not vertical except when the relative wind is horizontal.

除了相对风是水平的情况之外，升力不是垂直的。

6）This method can also be applied to retreat longwall panels with multiple entries, except that the holes will have to be drilled from the second entry.

除了钻孔必须从第二条平巷钻起外，这种方法也可用于具有多条平巷的后退式长壁采煤盘。

此句还可译为：这种方法也可用于具有多条平巷的后退式长壁采煤盘区，只是钻孔必须从第二条平巷钻起。

7）The construction and the working principles of the driven return end are similar to those of the drive head except that the return end does not require the minimum height for unloading.

除了回链端对于卸载的最低高度没有要求之外，驱动的回链端在结构和工作原理上与机头相同。

8）Longwall retreat mining is basically the same as longwall advancing extraction except that the coal seam is block-out and then retreated in panels between development roadways.

长壁后退开采基本上与工作面推进开采相同，只不过是煤层被阻挡，然后在开拓巷道间区段中后退开采。

以上句中 except 是介词，后面 that 连接宾语从句，可译为“除……之外”，或“不同的是……”。

（三）表语从句的翻译

在句子中充当表语成分的句子叫作表语从句。它是由 that、what、why、how、when、where、whether 等连词和关联词引导的。一般来讲，可以先译主句，后译从句。翻译表语从句一般都可以按原文顺序翻译，即先译主句后译从句，多采用“是”的句式，但也可以灵活处理。

1. 按原文的顺序翻译

1) The first reason of no pump pressure may be that the absorbent filter is blocked.

泵不上压故障产生的第一个原因可能是吸液过滤器堵塞。

句中主语是 the first reason，may be 后面的 that 引导的从句说明 the reason 的特征，所以是表语从句。

2) The general concept about longwall mining is that since there is a complete extraction of coal in a longwall panel, there will be much larger surface subsidence and thus structural damages.

长壁开采的一般概念是，由于在长壁开采的区段中煤被全部采空，会有较大范围的地表下沉，从而造成地面建筑物破坏。

句中 that 引导的是表语从句，句子主干是 there will be ... structural damages，说明 concept 的特征；后面 since 引导的原因状语从句说明原因。

3) The essential property of an explosives is that it can be converted very rapidly into gases which occupy many times its original volume, and when confined as in a shot-hole, these gases exert enormous pressure, sufficient to disrupt or break up the surrounding strata.

炸药的重要性质是能够迅速转化为是原体积许多倍的气体，一旦受到限制，如在炮眼内，这些气体便会释放出巨大的压力，足以使围岩碎裂或破落下来。

本句结构比较复杂，is 后面是 that 引导的表语从句，由两个以 and 连接的并列句构成，其中还有定语从句 which occupy many times its original volume 修饰前面的 gases，和时间状语从句 when confined as in a shot-hole 表示时间。

2. 表语从句几种常见句型的翻译

在 that/this is why...句型中，如果选择先译主句，后译从句，可以译成“这就是为什么……”“这就是……的原因”“这就是……的缘故”等。如果选择先译从句，再译主句，一般可以译为“……原因就在这里”“……理由就在这里”等。在 this/it is because...句型中，一般先译主句，再译从句，译成“是因为……”“这是……的缘故”。例如：

1) Different methods of coal cutting produce different amounts of methane emission, and this is why coal seams with high methane content are not suitable for hydraulic mining.

不同的煤炭切割方法会产生不同的甲烷排放量，这就是为什么甲烷成分高的煤层不适于液压开采。

2）The reason is that the total electricity production in Spain kept rising.

西班牙的电力总产量持续上升，原因就是如此。

3）This is because the overburden was undisturbed，and the water injected into the borehole primarily flowed in the pre-existing fractures.

这是因为覆岩未受采动影响，注入的水主要流到了覆岩的原生裂隙中。

4）This may be because low strength coal tends to deform in and around the face area as mining occurs，resulting in stress relaxation.

这可能是因为采矿时，低强度煤层易在采矿工作面或其周围发生变形，导致应力释放。

（四）同位语从句的翻译

同位语是用来进一步解释名词或代词的，同位语可以是单词、短语也可以是从句。同位语从句一般由 that 引导，但也可以用关系代词 what、which、who、关系副词 when、where、why、how 或 whether、if 引导。同位语从句常常解释和说明的名词主要有 fact、thought、theory、idea、hope、news、doubt、conclusion、question、problem、evidence、certainty、belief、rumor、mystery、suggestion、order、answer、decision、discovery、explanation、information、knowledge、law、opinion、principle、truth、promise、report、statement、message、saying、rule 等。同位语从句经常容易跟定语从句混淆，它们的不同在于：同位语从句和所说明的名词是补充说明关系，说明一个名词的具体内容；定语从句则形容一个名词的性质和特征，是修饰性的，作所修饰名词的定语。同位语从句的翻译方法主要有以下几种。

1. 顺译法

1）The greatest problem lies in the fact that most of the energy input to a crushing or grinding machine is absorbed by the machine itself，and only a small fraction of the total energy is available for breaking the material.

最大的问题在于输入破碎机或磨碎机的大部分能量都被设备本身吸收了，真正用于碎裂物料的能量只占总输入能量的很少一部分。

句中 that 引导的从句是补充说明前面的 fact，说明它的具体内容，所以是同位语从句。

2）A problem connected with some chainless haulage systems is the fact that they impede the flexibility of the face conveyor and can cause operational restrictions.

某些无链牵引系统存在的一个问题是，它们妨碍工作面输送机的弯曲性能，可能使运行受到约束。

3) There is strong evidence that certain coal and lignitic deposits were formed by material that was transported from other areas.

有令人信服的证据说明：某些煤层和褐煤层，是由其他地区运输来的材料形成的。

句中有两个 that 引导的句子，但是语法作用各不相同：第一个 that 连接的是同位语从句，说明主句的主语 evidence、that 作为引导词不能省略，在从句中不作语法成分；第二个 that 连接的是定语从句，修饰 material、that 在定语从句中作主语。

4) These facts leave no doubt that coal is made up of plant material, but to understand how Britain's coal seams were formed over 200 million years ago, we must first study something that is happening at the present day.

这些事实无疑地说明煤是植物物质组成的，但要了解两亿多年以前英国的煤层是如何形成的，我们必须首先研究一下目前正在发生的一些情况。

本句的主干是 these facts leave no doubt，句中 that 引导的从句是说明 facts 的内容，所以是同位语从句，为了避免头重脚轻的结构，同位语从句放在了后面。

5) This is due to the facts that it serves as the major exit for the high-concentration methane in the gob.

原因在于，它是采空区高浓度沼气的主要泄出口。

2. 译为类似定语的结构或独成一句

1) This is due to the fact that in the sealed gob, there is insufficient amount of oxygen in the air.

这是由于在封闭的采空区内，空气中氧气含量不足。

句中介词短语 due to，通常译为“由于……”，that 引导的句子说明 fact 的具体内容，所以译成定语成分更加自然。

2) The fact that the gravity of the earth pulls everything towards the center of the earth explains many things.

地球引力把一切东西都吸向地心，这一事实解释了许多现象。

3. 使用冒号、破折号或“这样”“这一”“即”等词语

1) The industry has already accepted the fact that the majority of the world's future minerals will come from low-grade, super-large, high-tonnage, and ultra-mechanized operation.

该行业已经接受了这一事实：未来世界上大部分的矿产将来自低品位、超大型、高吨位和超机械化作业。

2) Ordinary concrete is not an ideal material for mine supports, for the simple reason that while it offers considerable resistance to crushing or compression, it is not capable of withstanding tensile stresses which may arise from ground movement.

普通混凝土并不是矿井支护的理想材料，理由很简单，即它虽然有相当大的抗碎及抗压能力，但不能经受地层移动而引起的张应力。

3) And there was the possibility that a small electrical spark might accidentally bypass the most carefully planned circuit.

而且总有这种可能性：一个小小的电火花，可能会意外地绕过了最为精心设计的线路。

二、定语从句的翻译

矿业英语中，定语从句的使用比较多，所以对定语从句的翻译要给予足够的重视。在英语中，用来修饰、限制、说明句中某一名词、代词、名词短语或代词短语乃至整个句子的从句叫定语从句。定语从句很常见，有的结构比较简单，有的结构相当复杂，有的与先行词关系密切(限制性)，有的与先行词关系不怎么密切(非限制性)，有的定语从句还具有原因、结果、让步、目的、条件、假设等意义。翻译时有比较大的灵活性，可根据其结构和含义采用不同的翻译方法。

定语从句分为限制性定语从句与非限制性定语从句两种，它们在英语中的位置一般是在其所修饰的先行词后面。限制性定语从句与非限制性定语从句的区别在于限制意义的强弱，对它们采用的翻译方法也不尽相同。

(一) 限制性定语从句的翻译

1. 前置法

限制性定语从句与先行词关系密切，尤其是一些较短的限制性定语从句，没有它主句的意义便不完整，可按照汉语定语前置的习惯将其翻译成带“的”的定语，放在先行词前面。例如：

1) Double-acting reciprocating compressors are machines in which compression is effected in both ends of the cylinder.

复动式往复式压缩机就是在汽缸两端都压缩空气的机械。

2) After the treatment sewage mixed with the mine water is discharged into the reservoir that is formed by subsidence area.

污水经处理后，连同矿井水混合排入利用塌陷区所形成的水塘。

3) This mining method still finds occasional use in mining high-grade ores or in countries where labor costs are low.

这种采矿法仍然常会应用在高品位矿体开采或劳动力成本低的国家。

4) Stull stoping can be applied to ore bodies that have dips between 10° and 45°.

横撑支柱采矿法可用于开采倾角为10°～45°的矿体。

5) Pit limits are the vertical and lateral extent to which the open pit mining may be economically conducted.

露天开采境界为露天矿可以进行经济合理开采的垂直方向与水平方向的范围。

以上各句中的限制性定语从句与所修饰的词关系密切，采用定语前置的方法能对所修饰词起更好的限定作用。

2. 后置法

上述译成前置定语的方法大都适用于限制性定语从句，但一般用于译比较简单的英语定语从句；当定语从句较长时，如果翻译成前置的定语，就不太符合汉语的表达习惯，在这种情况下，往往把该定语从句翻译成并列的分句，放置于原来它所修饰词的后面。在处理此类定语从句时，一般遵循的原则是：译成并列分句，重复英语先行词或译成并列分句，省略先行词。例如：

1) Continuous mining involves the use of a single machine known as a continuous miner that breaks the coal mechanically and loads it for transport.

连续开采就要提到一种机器，叫作连续采煤机。它机械地破煤，并且为运输装载煤。

2) Block caving is most applicable to weak or moderately strong ore bodies that readily break up when caved.

矿块崩落法最适合于稳固性差或者中等稳固矿体的开采，这样的矿体崩落后能很容易形成碎块。

3) There is the surface map which shows the location of topographical features such as coal and other mineral outcrops, streams and rivers, roads and highways, railroads, core drill holes, oil and gas wells, and the boundary lines of the mineral property itself.

一类是地面图，它表示地貌特征，诸如煤及其他矿物露头、河流、道路、铁路、钻孔、油气井的位置以及矿藏自身的边界线等。

4) During mining operations, the coal seams and the surrounding strata are subjected to continuous fracturing which increases the passageways for the

methane and destroys the equilibrium between the free and adsorbed methane that exists under natural conditions.

在采煤作业过程中，煤层和围岩会遭到连续的断裂，这就给沼气增加了通道，打破了天然状态下游离沼气与吸附沼气之间的平衡。

以上四个例句中的定语从句都比较长，如译成汉语中的前置定语则显累赘，故采用了重复先行词的方法，为了避免重复，也可用代词“它”来代替。

5) The boreholes are filled with explosives which break up the rock when exploded.

这些炮眼里装满炸药，爆炸时把岩石炸碎。

句中 which 引出的定语从句，修饰前面的名词 explosives，which 在从句中作主语，同时从句中还有省略了主语的时间状语从句。

6) The underground water which, being dirty and sometimes in large quantities, is brought there from the underground districts is continuously pumped out of the mine.

这种地下水，已被污染而且有时量很大，从井下各区域汇集到这里，然后再持续不断地被排出井外。

7) There are other unwanted gases which it is the job of the ventilation system to carry away.

矿井中还有些其他有害气体，而通风系统的工作就是排出这些气体。

本句中 which 引导定语从句，修饰名词 gases，而 which 在从句中作不定式短语 to carry away 的宾语。

3. 融合法

融合法是指把原句中的主语和定语从句融合在一起译成一个独立句的翻译方法。因为限制性定语从句与主句关系紧密，所以融合法比较适用于翻译限制性定语从句。

通常的做法是把原句中的主语和定语从句融合，把主句压缩成词组译作主语，而把定语从句的动词译作其谓语。例如：

1) There are a number of recent publications which provide considerable information on estimating costs for various sizes and types of mines.

最近许多出版物提供了不少有关不同规模和类型矿井费用预算的资料。

翻译本句时，主句和定语从句融合后，把 publications 译成主语，把定语从句中的动词 provide 译成谓语，使整个译文更加精炼通顺。

2) A code is a set of specifications and standards that control important technical specifications of design and construction.

一套规范和标准可以控制设计和施工的许多重要技术环节。

3) Gathering pumps are those which pump the water from its source either into the main sump or into an intermediate sump.

汇流水泵将水从水源泵导至主水仓或中间水仓。

以上例句在翻译过程中,都是采用了融合法,把主语和定语从句融合,主句压缩成整句的主语,从句的谓语译成整句的谓语,使得译文更加简洁流畅。

（二）非限制性定语从句的翻译

非限制性定语从句同其先行词之间的联系是松散的。它不是句中必不可少的组成部分,而仅是对先行词作些描写或补充说明。非限制性定语从句前常有逗号将它与主语分开。翻译这类从句可以运用下面几种方法。

1. 前置法

一些较短而具有描写性的英语非限制性定语从句,也可以译成带"的"的前置定语,放在被修饰词前面;但这种处理方法不如用在英语限制性定语从句中那样普遍。

1) Colliery explosions may be explosions of firedamp, explosions of coal dust or mixed explosions, in which both these agents play a part.

煤矿爆炸可以是瓦斯爆炸、煤尘爆炸或者是两者兼有的混合爆炸。

2) Basic equipment and methods, which can be combined in various ways, depending upon conditions, for efficient drifting, are summarized in the following.

为了高效率开掘平巷,依据条件,把能以各种方式组合的基本设备和方法归纳如下。

以上例句中,which 引导的都是非限制性定语从句,但相对内容较短,这种情况,可以译成简短的前置定语。

2. 后置法

英语的定语从句结构常常比较复杂,如果翻译在其修饰的先行词前面的话,会显得定语太臃肿,而无法叙述清楚。这时,可以把定语从句翻译在先行词后面,译成并列分句。翻译时可以通过重复先行词或省略先行词这两种方法来处理。

(1) 重复先行词。

由于定语从句的先行词通常在定语从句中充当句子成分,如果单独把定语从句翻译出来的话,常常需要重复先行词,还可以用代词代替先行词。需要注意的是非限制性定语从句修饰先行词或整个主句,起解释、补充或附加说明作用,如果省略掉,主句句意仍保持完整。例如:

1）The forest swamp condition lasted for many centuries, during which the trees as they died formed a thick sludge of partly decayed vegetable matter, giving a kind of peat.

森林沼泽地这一状态延续了很多世纪，在此期间，死掉的树木形成了含有部分腐殖质的一层很厚的淤泥，从而生成了泥炭。

本句中，which 引导非限制性定语从句，修饰名词 centuries，由介词 during 引导，翻译时可以用代词代替先行词，译为“在此期间”；该定语从句中还包含一个时间状语从句 as they died。

2）A dome is an opposite type of structure to a basin, which is characterized by a radial arrangement of directions away from a central area or point.

穹窿是一种与盆地相反的构造，它的特点是从中心区域或中心点向外倾向呈放射形排列。

本句中，which 引导的非限制性定语从句修饰整个句子，短语 be characterized by 意思是“以……为特点”，翻译中也采用了用代词代替先行词的方法。

3）At the surface, the outer and inner pipes are connected respectively to two horizontal distributing rings, which in turn are connected with a pump and ice-machine。

在地面，外管和内管分别连接到两个水平分配环上，这两个分配环依次和水泵、冷冻机连接。

本句中，which 引导的非限制性定语从句修饰先行词 two horizontal distributing rings，翻译时，重复了先行词“这两个分配环”，使得语义更加明确。

4）Graders and water trucks are absolutely essential to haul road maintenance, which in turn is one of the most important elements of an efficiently operating surface mine.

平地机和运水车对运输道路养护是绝对必要的，而公路养护又是高效运行露天矿的最重要因素之一。

5）The incoming supply to the winder is controlled by an oil circuit-breaker, which must provide short circuit protection for the equipment.

提升机的输入线的控制用油断路器，它应对整个设备提供短路保护。

（2）省略先行词。

如果把定语从句翻译在先行词后面，在通顺、完整的前提下，有的时候也可以不用重复先行词。例如：

1）Coal，which is the main resource for China's energy supply，plays an increasingly important role in national energy consumption.

煤炭是我国的主要能源，在国家能源消费中占据了很重要的位置。

2）There are a variety of drilling patterns，which are individually designed for the size of the excavation，the type of rock at the face，and the drifting technology applied.

凿岩布置方式有多种，主要根据井巷的大小，工作面岩石类型，使用的平巷掘进技术等进行具体的设计。

3）Fan characteristics are a matter of design，which is controlled by the manufacturer.

风机的特性取决于设计，由制造商掌握。

以上例句的译文都省略了先行词，使译文结构更加紧凑，符合汉语的行文习惯。

（三）翻译成状语从句

虽然从语法结构上讲定语从句属于修饰成分，修饰前面的先行词，但是从功能上来讲，有些定语从句兼有状语从句的职能，说明原因、结果、目的、让步、假设等关系。翻译时应善于从原文的字里行间发现这些逻辑上的关系，然后译成汉语各种相应的偏正复句。例如：

1）Good track is also essential to eliminate derailments，which are a common source of accidents.

脱轨是一种常见的事故源，因此高质量的轨道对消除脱轨事故很关键。

本句中，主语和从句之间是有因果的逻辑关系，这时，译为表原因的偏正复合句能更准确地表达原句的意思。

2）Automatic machines，which have many advantages，can only do the jobs they have been told to do.

虽然自动化机器有很多优点，但它们只能做人们吩咐它们要做的事。

本句中，从句表达的是让步的逻辑关系，这时，译为表让步的偏正复合句能更准确地表达原句的意思。

3）Slow delays are used primarily underground and in tunnel work，where they provide sufficient time for rock movement between delay periods.

长延时雷管主要用于井下和隧道掘进时的爆破，目的是使岩石在延时内有足够的时间移动离开。

本句中，从句表达的逻辑关系更趋同于主语的目的，故可以直译出来。

4）There are many types of centrifugal mine fans，which differ in detail of design.

有很多种矿用离心式通风机，但是其结构细节各不相同。

根据主句和定语从句之间的逻辑关系，以上例句中的定语从句分别译成了表示原因、让步、目的、转折的偏正复合句，所以在翻译时理清主句和定语从句之间的逻辑关系非常重要。

三、状语从句的翻译

矿业英语中的状语从句内容丰富，涉及面广，可以用来表示时间、原因、条件、让步、目的、地点等意义。英语中状语从句的翻译，一般比较容易处理，通常可以直接翻译。需要注意的是在汉语译文中，要如何将状语从句置于恰当的位置，并如何将其与主句之间自然连贯地连接起来，如何按汉语的习惯表达将句子类型进行相应的变化。由于两种语言表达上的不同，在状语从句的安排方面，存在着明显的差异，所以在翻译的时候也需要根据汉语习惯来灵活翻译。

（一）时间状语从句

1. 译成相应的时间状语

在英语中，时间状语从句的位置很灵活，可以放在句首、句末，甚至句子中间。但汉语的行文习惯通常是将时间状语放在句首，所以翻译的时候原则上应将时间状语译在句首。例如：

1）As the mine workings advance, various connections between the intake and return airways must be sealed.

当井下巷道前移时，进风巷与回风巷之间的各个通道必须封闭。

2）When the methane content in fresh air is 9.5%, once it encounters a heat source of sufficient temperature, the whole amount of methane and oxygen will participate in the chemical reactions.

当新鲜空气中沼气的含量达到9.5%时，一旦接触具有足够温度的热源，全部沼气和氧气便一起参与化学反应。

需要注意的是，本句中的 when 和 once 都是连词，分别连接两个时间状语从句。

3）When using permissible explosives, as long as the shot-firing is properly implemented the methane will not be ignited.

在使用安全炸药时，只要爆破作业进行得当，沼气是不会爆炸的。

when using…分词短语作时间状语，在句中强调其时间意义。

以上例句的翻译中，都是采用了时间状语放在句首的方法，比较符合汉语的行文习惯。

2. 译成汉语的并列句或平行结构

1) Once the manways, ventilation raises, and service ways have been established for a stope, mining can commence.

一旦采场的人行材料天井、通风井和辅助巷道形成后,就可以开始进行矿石回采了。

2)As the slope angle is increased, the number, size, and movement rate of slope failures increases.

边坡角增大后,边坡垮塌的次数、规模、移动速度就会增加。

3. 译成条件句

由于时间状语的引导词除了显示时间关系之外,有时候可以表示条件关系,所以还可以翻译为条件句。例如:

1) When outbursts occur, they can be very serious events, possibly even resulting in multiple fatalities.

如果发生煤岩突出,那将是很严重的事件,甚至可能导致多人死亡。

2) The aim of gas drainage is to lower the gas content of the seam below a certain threshold value, at which time it is considered safe to mine the seam.

瓦斯排放的目的是将煤层瓦斯含量降低到某一阈值以下,只有这样,对煤层的开采才会安全。

3) We can't start the job until we have the approval from the authority concerned.

如果没有有关当局的批准,我们不能开始这项工作。

(二) 原因状语从句

矿业英语中,原因状语从句的连接词常常有 because、since、as、now that、seeing that、considering that、in that、in view of the fact that。在翻译的时候,大多数原因状语从句可以放在主句之前。

1. 译成表示原因的分句,放在主句之前翻译,显示前因后果的关系

1) Since the specific gravity of methane is very small, it tends to accumulate near the roofline and forms a methane layer, sometimes up to 200-300 mm thick.

由于沼气比重很小,所以一般积聚在靠近顶板处,并形成一个沼气层,其厚度有时可达到 200～300 毫米。

2) Cushion blasting has been done to depths near 30 m in a single lift with the larger-diameter boreholes because alignment is more easily retained.

由于采用大直径钻孔,钻孔方向容易控制,所以缓冲爆破一次爆破的高度可

达 30 米。

3) There would be almost no catch benches left, as the bench face angle would be steeper than the bedding.

由于台阶坡面角大于岩层倾角，基本无法布置安全台阶。

以上例句，在翻译过程中，都把原因分句放在了句首，比较符合汉语中先因后果的行文习惯。

2. 译成因果偏正复句中的主句

1) Since a pit geometry is required to define design sectors, slope design is interactive with mine planning.

确定设计区段需要露天采坑的集合参数，因此边坡设计是一个渐近过程，与开采计划相互作用。

2) Almost all the hard rock mines now have circular shafts because the cross section provides good geometry for airflow and good rock support characteristics.

目前几乎所有的坚硬岩石矿井均已经采用圆形井筒，因为这种断面形状通风效果好，而且这种断面岩石支护效果好。

（三）条件状语从句

条件状语从句在翻译的时候，一般可以译成表“条件”或“假设”的分句，放在主句之前或主句后面，有时候，还可以根据上下文省略连接词。有的可以译为表补充说明情况的分句，放在主句后面。例如：

1) If the stress exceeds the strength, the slope is unstable; conversely, if the strength exceeds the stress, the slope is stable.

如果应力超过强度，边坡就不稳固。反之，如果强度大于应力，边坡就稳固。

2) Considerable savings in freight cost will accrue if the smelter is located far from the mine.

如果冶炼厂远离矿山，节约的运输成本累积起来也相当可观。

3) If it employs a solvent that can be recycled, the process may be cheap enough to be applied to very low-grade ores.

如果使用的是可以循环再生的溶剂，这个方法成本就非常低，可用于回收品位极低的矿石。

以上例句中的条件状语从句，都译成了表“假设”的分句，放在了句首；条件从句在译文中的具体位置并不固定，应当根据整句的逻辑关系来决定译文中条件从句的位置，这样才能准确传达语义，译文通顺自然。

4) Very flat shafts can be sunk at speeds little less than for driving

tunnels, unless there is much water.

倾角非常小的井筒就能以略小于平硐的速度掘进,如果没有很多水的话。

5) Iron or steel parts will rust, if they are unprotected.

铁件或钢件是会生锈的,如果不加保护的话。

以上例句中的条件状语从句在译文中都位于主语的后面,起到补充说明的作用,同时突出了主句的重要性。

(四)让步状语从句

1. 译成表示"让步"的分句

1) Although peat will burn when dried it has a low carbon and high moisture content relative to coal.

尽管干燥后的泥炭也可以燃烧,但相对于煤来说,它的碳含量较低,而水分含量较高。

2) Although coal forms less than one percent of the sedimentary rock record, it is of foremost importance to the Bible-believing geologist.

尽管煤炭在沉积岩记录中所占比例不到1% ,但对于相信《圣经》的地质学家来说,煤炭是最重要的。

2. 译成表"无条件"的条件分句

1) No matter what the shape of a magnet may be, it can attract iron and steel.

不论磁铁形状如何,它都能吸引钢铁。

2) He got the same result whichever way he did the experiment.

不论用什么方法做实验,他所得到的结果都相同。

(五)目的状语从句

汉语里表"目的"的分句所常用的关联词有"为了""省得、免得""以免""以便""生怕",其中"为了"常用于前置分句,"省得、免得""以免""以便"等一般用于后置分句。

1. 译成表示"目的"的前置分句

1) The pump room should be well lighted in order that all parts of the pump may be easily seen.

为了容易看清水泵的所有零件,水泵房应该光线明亮。

2) Measures should be taken to minimize the ground pressure behavior so that these behaviors do not interrupt mining operations and threaten the safety of coal miners.

为使矿压显现不致影响采矿工作正常进行和保障安全生产,必须采取各种技术措施把矿山压力显现控制在一定范围内。

2. 译成表示“目的”的后置分句

1) The Ventilation Regulations require a water gauge to be kept in every fan house so that a check can be kept on the difference in pressure developed by the fan.

通风规程要求每个通风机房里都要有一个水柱负压计，以便检查出由通风机产生的压差。

2) Ample space should be left around reciprocating pumps so that any pistons or rods can be pulled straight out from the pump.

往复泵周围应留有宽敞的空间，以便活塞或活塞杆能从水泵中直接拉出。

（六）地点状语从句

矿业英语的地点状语从句可以顺译为汉语中的地点状语，但更常译成条件状语，具体方法的选择要视语境而定，使译文准确、通顺、流畅。例如：

1) Where it is not practicable to effectively use water sprays to keep down the bulk of dust, the ventilation system must clear away the dust from places where men are working.

在那些不能有效使用喷水法控制大部分矿尘的地方，通风系统必须从人们工作地点将矿尘排除掉。

本句中有两个从句，第一个是 where 引导的地点状语从句说明地点，第二个 where men are working 是作定语来修饰前面的名词 places。

2) In any case, a hung-up stope is a costly and dangerous problem, and shrinkage stoping should not generally be used where the ore has a tendency to hang up.

任何情况下，采场出现堵塞都是一个伤财且危险的问题，因此，对于矿石有结块堵塞倾向的矿体不应选用留矿法。

句中 where 引导的是地点状语从句，表示留矿法适用的地方。

第八节　被动语态的翻译

英汉两种语言都有被动形式，但使用情况却很不相同。从使用范围上来说，被动语态在英语表达中使用非常广泛。通常在不必或不愿说出施动者，无法指出施动者，强调受动者，或者是为了使上下文更流畅等情况下，英语都会使用被动语态。汉语中由于动词本身可表达被动含义，因此虽然也有被动语态，却很少使用。从英汉语言结构上看，英语中被动语态是用动词的分词形式来构成的，即

“助动词＋过去分词”，汉语动词则没有这种形式。此外，被动语态是文字表达客观化的手段之一，可以增强论述的客观性，因而被动语态在英语科技文体中使用得尤其广泛，而汉语这一特征则不明显。由于这些差异，在被动句的翻译中，不能够生硬地一一对应，需要我们按照译入语的习惯，灵活地选择适当的句式。在英译汉的翻译过程中，大部分被动语态的句子都可以译成主动句。而在汉译英的过程中，许多汉语主动形式的句子也可以按照英文表达习惯调整为被动句。

一、英译汉中被动语态的处理

（一）译成汉语被动句

汉语中被动句式虽然使用范围小，但是也有用被动形式来表达的情况。这一类句子都是强调被动的动作，句子可以说出动作的施动者，也可以不说。一般来说，英语原句中若着重指出行动被施加在受动者身上的事实，在翻译时都可译为汉语的被动句。被动语态在英语中使用范围颇广，但其被动标记却很单一，主要是“be 动词＋动词过去分词”的形式。而汉语被动句式虽使用较少，其被动标记却比较繁杂。因此英语被动句在翻译为汉语被动句时，除了用“被……”这样典型的被动标志词语以外，还可以灵活地采用“由……”“受……”“受到……”“挨……”“遭受……”“为……所……”等结构。

1. 译为“被……”的形式

在强调被动动作时，句子可以译成汉语的被动句，即含有“被……”的句式，这时对原句的结构顺序没有太大的变动。例如：

1) The period of peat development varies from several hundred to several thousand years, and the transformation from peat to coal is considered a geologic age.

泥炭发展的周期从数百年到数千年不等，而从泥炭到煤层的转化过程被认为是一个地质时代。

原文中的 is considered 直接译成“被认为”，原句的被动句式和主语都没有改变。

2) Up to now, sulfur dioxide has been regarded as one of the most serious of these pollutants.

到目前为止，二氧化硫一直被看作是这些污染物中最严重的一种。

has been regarded 译为“一直被看作”保留了被动形式，符合汉语的表达习惯。

3) Peat or sapropel, after being buried, transforms into lignite under the influence of multi-factor.

泥炭或腐泥被掩埋后，在多种因素影响下，转变成为褐煤。

4) Coal mine backfill, as a validated technique, has been used in many areas in China.

充填开采技术已经成熟，被应用在国内许多矿区。

5) A safety ladder is used for lifting personnel safely in an emergency situation, and is suspended above the working face of the shaft during shaft sinking.

凿井时安全梯被悬吊于井筒工作面上方，在紧急情况下供人员安全升井。

6) The core zone occupies the center portion of the pillar and is defined as the part of pillar that does not experience any plastic deformation.

核心区位于煤柱的中心，被定义为没有经历任何塑形变形的煤柱部分。

7) The mine-out area in room and pillar mining is called "room-and-pillar mining gob".

房柱式采煤法采煤后留下的采空区被称为"房柱式采空区"。

8) Heat is regarded as a form of energy.

热被看作能量的一种形式。

2. 译为"受……""让……""由……""遭……""得到……""为……所……"等形式

按照汉语的表达习惯，有些汉语被动句用"被……"来引导会显得不通顺，这时可以用其他一些可以表示被动的词"受、让"等来替换，原句的结构同样也不需有太大变化。例如：

1) The seams are shallow, thick, nearly horizontal, and generally not affected by geological tectonic activities.

该煤层资源埋藏浅，煤层厚，煤层倾角近乎水平，且基本上不受地质构造影响。

原句中的 are not affected by 部分的翻译，保留了原文的结构，按照汉语习惯翻译成了"不受影响"。

2) The long-standing problems such as a low recovery rate, accumulation of coal bed methane and coal dust have never been solved.

长期存在的回采率低，煤尘与瓦斯聚集等问题一直未能得到解决。

3) Gradient of coal metamorphism is expressed by the quantity in the reduction of volatile matters or increase in vitrinite reflection.

煤变质梯度常由挥发分减少或镜质组反射率增高数值来表示。

4) Over the years, tools and technology themselves as a source of

fundamental innovation have largely been ignored by historians of science.

长久以来,工具和技术本身作为根本性创新的源泉在很大程度上为科学史学家们所忽视。

5) A magazine must be approved by authorities and designed and built according to special provision.

爆炸材料库需得到主管部门批准,按专门规定设计建造。

6) Integration of coal resources is only a post-event management and shall be replaced by a market access system which is a pre-event regulatory measure.

煤炭资源的整合只是事后处理,应由市场准入制度取代,这是事前监管措施。

7) The maximum horizontal stress theory was proposed by W. J. Gale in the early 1990s.

最大水平应力理论是由 W. J. 葛尔于 20 世纪 90 年代初提出的。

(二) 译成汉语主动句

由于在汉语中被动形式使用较少,许多英语被动句在翻译时都可以按汉语表达习惯翻译为汉语主动句。英语中常用受动者作为句子的主语,以此作为谈论的主题。而将英语被动句译成汉语主动句时,译者应该按汉语表达习惯选择主语。主语可以是原句的主语,也可以是 by 后面的名词或代词,也可以是原句中隐藏的动作执行者。

1. 原句主语仍作主语

当英语被动句中的主语为无生命的名词,而且句中不出现由 by 引导的行为主体时,翻译时仍然将原句中的主语译成主语。例如:

1) In other words, mineral substances must be extracted by digging, boring holes, artificial explosions, or similar operations.

换言之,矿物质必须经过挖掘、钻孔、人工爆破或类似的作业才能获取。

此处译文利用了汉语的特点,用“获取”一词的主动形式表达被动含义,没有再加上被动标示词,使译文简洁通顺,更符合汉语表达习惯。

2) Coal rank can be used to represent the degree of coalification.

煤阶可用来表示煤化作用深浅程度的等级。

这个例子和上一句一样,用“可用来”这个主动形式的词表达被动含义,更符合汉语表达习惯。

3) For the steeply dipping faces, the gravity weight of coal cannot be ignored due to the angle influence.

由于角度影响，大倾角煤层的煤体自重作用不可忽略。

4）The supporting method was first adopted in the 2nd Mining District.

这种支护方法首先在第二采区得以应用。

5）The geophysical techniques can be applied to the investigation of coal deposits and the improvement in controlling the coal mines surrounding rock.

地球物理技术可用于煤矿床调查和改进煤矿围岩控制。

6）The geological radar method can be used to detect faults and solve hydrogeological and engineering geological problems.

地质雷达法可用于探测断层，解决水文地质、工程地质问题。

7）The skip loading pocket chamber is equipped with loading devices that can automatically load coal, ore or waste rocks into the skips.

箕斗装载硐室安装有将煤炭、矿石或矸石自动装入提升箕斗的装载设备。

8）Lagging is installed around the supports for transmitting ground pressure to the supports uniformly and prevent the rock from caving in.

背板安设在支架外围，使地压均匀传递到支架并防止碎石掉落。

2. 使用 by 后面的施动者作主语

有时英语被动句中的施动者以 by 结构引导出来，这时可采用颠倒顺序的译法，将 by 结构引导的部分译成句子的主语。例如：

1）When the face is approaching the fault plane, the rock pressure and roof movement are only affected slightly by the fault.

当工作面推进到断层面附近，断层对矿压显现和顶板活动影响不大。

原句在翻译为主动句时，由于 by 结构已指出了实际主语，按照汉语表达习惯将实际的施动者“断层”作为句子的主语。

2）A multi-seam interaction analysis software program was developed by scholars to determine the amount of damage in the upper seam when the lower seam was mined.

为了分析下煤层开采对上煤层开采的定量影响，学者们开发了多煤层相互作用分析软件。

was developed 表达的是“软件被学者们开发了出来”，译文按照汉语习惯，把实际施动者作为句子的主语，翻译成“学者们开发了软件”。

3）Even when the pressure stays the same, great changes in air density are caused by changes in temperature.

即使压力不变，气温的变化也能引起空气密度的巨大变化。

4）A new kind of substance has been found by scientific workers.

科学工作者已经发现了一种新物质。

5) Numerous empirical equations have been proposed by scholars to determine the pillar strength.

学者们提出了诸多经验公式用以计算煤柱强度。

6) A simplified model of pressure bulb theory for interaction between pillars was proposed by Syd Peng in 1980.

彭赐灯在1980年给出了煤柱之间压力球理论的简化模型。

7) The high stress or abutment zone at the longwall face and pillar ribs can be determined by the drilling yield tests.

钻屑法能够定位长壁工作面和煤柱的高应力区域或较高支承压力区域。

8) Only a small portion of solar energy is now being used by human beings.

现在人类只能利用一小部分太阳能。

3. 使用原句中隐藏的施动者作主语

有时英语被动句中并未出现by结构，而是包含构成句子状语成分的介词短语。这时可将介词短语中的名词或名词词组译成句子的主语。例如：

1) In this case, support should emphasize the entry corners, because both horizontal and shear stresses are concentrated on these areas.

在这样的情况下，支护重点应该放在巷道顶角，因为这些区域水平应力和剪应力集中。

原句的原因状语从句中包含了一个被动语态结构，译文把从句中的介词结构 on these areas 在翻译时调整为从句主语，使译文显得更通顺。

2) Surface subsidence observation stations were set up in several coal mines for surface subsidence measurement.

一些煤矿设置了地表观测站对矿区地表沉陷规律进行观测。

将介词结构状语 in several coal mines 在译文中变为句子主语，这样表达更加通顺。

3) The physical and mechanical properties of waste rock from seven different mining areas in China have been analyzed and assessed in this paper.

本文对取自中国七个不同地区的煤矸石试样的物理力学特性进行了分析和评价。

4) Based on the compaction characteristic of backfilling material, the thin plate roof model was built in this study.

本研究依据充填材料本身的压实特性，建立了地基薄板模型。

5）The construction plan, construction procedure and technical construction measures for shaft extension, shaft installation reconstruction, ground hoist and headframe replacement in auxiliary shaft extension project were introduced in this paper.

本文介绍了副井井筒延深工程中井筒延深、井筒装备改造、地表提升机及井架更换的施工方案、施工程序及施工的技术措施。

6）Features of the critical plane are discussed and the calculation of the damage parameter is presented in this model.

此模型分析了临界平面的特点并给出了损伤参量的计算过程。

（三）译成汉语无主句

在英语中，作者不知道、不想说或不必说出施动者时，便使用受动者作主语来表达思想。英语句子必须有主语，但汉语句子却可以没有主语。因此，在英语被动句的翻译中，如果原句不含动作的发出者，而且表示存在、观点和态度等时，常可以译成汉语的无主句。也就是把原文的“受动者＋动词＋（省略施动者）”变为“（省略施动者）＋动词＋受动者”的顺序。例如：

1）As a result, many different types of rib bolts have been developed.

由此形成了多种不同形式的煤壁锚杆支护。

本句描述了一个客观事实，原句也未指出施动者，汉语相应地翻译为无主句，这样比按原句顺序采用被动形式翻译要更为通顺。

2）Different gas drainage techniques were proposed for various stages of mining and the technique was demonstrated at Hebi Zhongtai Mining Co., Ltd.

针对不同的开采阶段提出了不同的瓦斯抽放模式，并将瓦斯抽放技术应用到鹤壁中泰矿业公司。

由于原文不必说出施动者，使用受动者为主语，汉语相应地翻译为无主句，符合汉语表达习惯。

3）Three physical models with slope angles of 20°, 30°and 45°were built.

分别建立了三个角度为 20°、30°和 45° 的物理模型。

4）Practically no scientific experiments have yet been made on the question.

对此问题实际上还没有进行过科学实验。

5）The sandstone percentage within 20 m above the coal seam is calculated and the effect of each rock type on the methane content of the underlying coal seams is discussed.

计算了砂岩厚度在煤层之上 20 米范围内所占的比例，讨论了各类型顶板对其下伏煤层甲烷含量的影响。

6) A hierarchical structure of self-organizing wireless sensor network for ground sound source location is presented.

提出一种分层结构的自组织无线传感器网络用于地下声音源的定位研究。

7) The roof lithology and its distribution characteristics are mainly analyzed.

着重分析了顶板岩性及组合特征。

8) The method of determining the index for predicting partial roof falls has been presented, and the possibility of using a comprehensive index for roof weighting prediction has been discussed.

给出了局部冒顶监报指标的确定方法，探讨了采用回采工作面来压综合判断指标来预报顶板来压的可能性。

(四) 译成汉语判断句

英语中说明时间、地点、方式、方法等具体情况，着重描述事物的性质和状态的被动句，常可以译成汉语的判断句，其结构是"是……的"。例如：

1) Surface subsidence prediction is used to evaluate possible subsidence on surface structures and the environment.

地表下沉预计是用来评价可能的下沉量对地表建筑物和环境的影响的。

原句说明了"地表下沉预计"的性质作用，翻译为汉语的判断结构，更符合汉语表达习惯。

2) These data were collected in Hecaogou Mine.

这些数据是在禾草沟煤矿收集的。

原句说明了事情的地点，翻译为判断句比采用被动形式通顺，且符合汉语表达习惯。

3) The failure mechanisms were analyzed by numerical modeling.

破坏机理是通过数值模拟手段来分析的。

4) In consideration of the prevailing conditions of coal reserves, China's coal resources are mainly mined by underground methods.

鉴于煤炭赋存条件的实际，中国煤炭是以地下开采为主的。

5) The resistivity method is used for solving geological problems by studying the resistivity variation of underground rock.

电阻率法是通过研究地下岩石电阻率变化来解决地质问题的。

6) Initial fissures are naturally formed in the process of rock formation.

原生裂隙是在岩体生成过程中自然形成的。

(五) it 作为形式主语的句式译为汉语的无主句

英语中有一类以 it 作为形式主语的句子，在科技文体的翻译中通常都可以改为汉语的无主句。下列就是一些常见的句型。

it is said/declared that… 据说、据称……

it is learned/described that… 据悉、据闻……

it is supposed/estimated/alleged that… 据推测、据估计……

it is reported/stated that… 据报道、据报告……

it must be pointed/admitted that… 必须指出、必须承认……

it will be seen/interpreted that… 可见、可以看出……

it cannot be denied that… 无可否认……

it has been proved/demonstrated that… 已经证明……

it may be confirmed that… 可以肯定……

二、汉译英中被动语态的处理

汉语中被动句使用较少，但是很多句子虽然没有“被”“给”等词语，逻辑上却存在被动关系。由于被动语态在英语中使用广泛，在汉译英时，汉语中的被动句以及其他一些含有被动关系的句子，都可以用英语的被动形式来处理。

(一) 原文中被动关系词的汉译英

汉语句子中有“被……”“给……”这类直接的被动关系词，或者包含“由……”“让……”“受……”“为……所……”“是……的”等结构，且显示出逻辑上的被动关系时，就可以翻译成英语的被动句。例如：

1) 进一步的离层产生的岩块形成半拱，在采空区被逐渐压实。

Further bed separation causes the blocks to form a semi-arch, and the waste rocks are gradually compacted on the gob floor.

原文含有被动关系词，译为 are gradually compacted 符合英文表达习惯，“被逐渐压实”。

2) 由于落后无序的开采，大面积的矿区遭到了破坏。

Due to their disorganized outdated mining method, large area of mines was damaged.

“遭到了破坏”呈现出被动关系，译为 was damaged，符合英文表达习惯。

3) 不同支架工作阻力下煤壁的破坏情况由图 8 所示。

The failure of coal face with different shield support resistance is shown in

Figure 8.

4）管子常会给空气中的氧腐蚀。

Pipes are often corroded by the oxygen in the air.

5）楔形破坏可能是由采掘活动或者流水润滑和潜在破坏面的摩擦阻力降低引起的。

Wedge failure could be triggered by mining activities or running water lubrication and the reduction in friction resistance of the potential failure planes.

（二）主谓语逻辑被动关系的汉译英

如果在汉语句子中没有上述表示被动关系的词汇，但是主谓语成分在逻辑上具有被动关系，汉译英时也可以译为被动结构。例如：

1）有效控制底鼓的方法大致分为两类。

Effective floor heave control methods are divided into two categories.

本句虽然没有出现被动词汇，但是汉语的动词“分为”用主动形式表达了被动含义，句子表达的是被动关系，因此英文也对应翻译成了被动句。

2）自从采用了新的技术，产量已经有了很大提高。

The production has been greatly increased since the application of the new technique.

原文“产量有了提高”，表示的是“产量被提高”，包含了逻辑上的被动关系，译文翻译为被动语态，更符合英文表达习惯。

3）大倾角大采高开采煤层后应力重新分配，在工作面煤壁前方的一定范围内形成超前支承压力影响区。

Stress is redistributed after the steeply dipping coal seam of large height is mined and the abutment pressure influence area is formed within a certain distance in front of the coal face.

4）37220 工作面采用一次采全厚放顶煤综合机械化开采。

The fully-mechanized full seam top coal caving mining method was employed in the 37220 panel.

5）模拟结果表明，随着锚杆长度增加，最大水平位移减少，锚杆长度控制在 1.5～2.0 倍开挖深度之间为宜。

Numerical simulation results show that the maximum horizontal displacement decreases with the increase of the bolt length, and the bolt length should preferably be controlled between 1.5 and 2.0 times of the excavation depth.

6）露天炸药只适用于地面或者露天矿爆破作业。

Opencast explosives can only be used in blasting operations on surface or surface mining.

7）内部隔离煤柱主要设计用于承受传递过来的支承载荷。

The internal barrier pillars are primarily designed to support the abutment load transferred over them.

8）从支架结构上讲，凡是有掩护梁的液压支架都统称为掩护式支架。

Structurally, all powered roof supports with an inclined caving shield at the back of the shield are called shields.

（三）无主句的汉译英

无主句是汉语中常见的句式之一，汉译英时这类句子大都可以译为英语的被动句。例如：

1）采用半数值半解析的方法建立了多参数非线性弹性地基薄板模型。

A multi-parameter nonlinear elastic foundation plate model was established using a semi-analytical and semi-numerical method.

在汉语中无主句没有主语，而英语中句子不能缺失主语，所以无主句常可以译为被动句。本句按英语习惯翻译成了被动句。

2）最后，在寺河煤矿进行了抽采实验，得到了井下煤气渗透率的具体数值。

Finally, field trials are undertaken in Sihe Coal Mine and the permeability of coalbed methane are calculated.

3）通过数值模拟预测多煤层布置出现的叠合应力，并对未开采区的顶板质量进行煤矿顶板分级。

The overlapping stresses in a multi-seam mining setting were analyzed using the numerical modeling, and the quality of a coal mine roof in unmined areas was graded.

4）通过现场实测、理论分析、数值模拟等手段系统地研究了煤体变形、破坏、片帮规律，顶板运动规律以及支架稳定性等问题。

The deformation, failure and rib fall of coal mass, movement characteristics of roofs and support stability are studied by employing the on-site measurement, theoretical analysis and numerical simulation.

5）结合损伤力学和能量守恒理论，推导出了顶板下沉量的理论公式。

Based on the damage mechanics and energy conservation theory, the roof vertical displacement equation is proposed.

6）在薄煤层且托伪顶开采上进行了生产实践，总结了一套顶板管理方法。

The mining of thin coal seams below a false roof is practiced, and the roof control method is summarized.

7）借助相似模拟实验，研究了顶板自然垮落的最小跨度、移动角和崩落角。

The minimal uncontrollable roof-fall span, angle of critical deformation and caving angle are studied by similarity simulation.

8）应该简单提一提这种新方法。

Brief reference should be made to the new method.

因为被动语态在英汉两种语言中使用情况不同，翻译时应灵活处理。一定要注意译文的表达习惯，照顾两种语言的差别。在英译汉时，尽量不要一一对应地使用“被”字来翻译，而是灵活地处理为不同句式。在汉译英时，也同样要注意这一问题，将汉语中具有逻辑上被动关系的句子按英文习惯灵活地处理为被动句式。

第五章　综合练习

一、句子翻译练习

（一）英译汉

（1）Solid backfilling mining has been proven to be an economically and environmentally feasible solution.

（2）The technique of tunneling by smooth blasting under complex geological conditions is presented in the paper.

（3）The similarity parameters of the simulation test are shown below.

（4）Therefore, the widely used probability integral method in China can be applied for subsidence prediction for the solid backfilling mining.

（5）In order to avoid causing instability to the dam during mining and improve the effective impounding water level, some measures were taken during and after mining.

（6）Based on the practical ground pressure control theory and physical simulation, the displacement of the overlying strata was studied.

（7）The saturated water absorption rate and the height of specimen are almost linearly related.

（8）The problems and challenges facing the simultaneous mining of coal and gas are pointed out and its future development trends are discussed.

（9）Based on the genetic types of depositional structure planes, the deformation and failure characters are analyzed and their mechanics characteristics and mechanics effects are studied through theoretical analysis and experimental investigation.

（10）It is obviously an advantage if it is possible to locate the shafts at the fairly central in order to equalize as possible the average length of the haulage and ventilation circuits, but in practice other considerations may outweigh these.

(11) Some of the most costly sinking in the past are those to be sunk through loose, running, water-bearing sand and weak rocks which always increase the cost enormously.

(12) Care is necessary, however, to ensure that the shaft is of adequate size, not only for winding the load of coal, but also to allow for ventilation.

(13) Modern mucking operations may utilize a new type of grab which consists of a boom, centrally-pivoted and traversing circumferentially round the perimeter of the shaft on a monorail fitted below the bottom deck of the stage.

(14) The complete operation of traversing radially and circumferentially together with the raising and lowering of the grab unit is accomplished by air-driven motors, and the control is carried out either from the bottom deck or from a cabin below the deck.

(15) In this process the wet ground is artificially frozen and then blasted and excavated as though it were solid rock.

(16) No substitutes, however, have the peculiar advantage of wood in failing gradually when loaded beyond its strength and in giving warning of approaching failure by audible cracking.

(17) Today, mining is often portrayed as a beast that destroys everything in its path and creates wealth for the few and ongoing misery for the many.

(18) In some cases no packs are built other than those at the gate or roadsides, and the roof is allowed to collapse fully in the waste.

(19) The longwall retreating system offers certain advantages as compared with the longwall advancing system, in that no gob roads are kept open and the goaf is left behind.

(20) The arrangement of the drums enables the whole seam to be cut in either direction of travel, thereby ensuring rapid face advance and shortening roof exposure time.

(21) If the powered support is not the immediate forward support (IFS) type, the unsupported area after cutting is larger, and is left exposed longer.

(22) So the knowledge that coal would burn and even some uses of that knowledge, go back thousands of years.

(23) The air-lock casing for the upcast shaft may be built inside the legs or alternatively it may enclose the legs.

(24) Increase in winding capacity is afforded by the use of multi-deck cages, i. e. cages with two or more decks, each deck carrying one or more tubs, depending upon the size of cage which can be accommodated in the shaft.

(25) Thus for the same output, a small load may be wound at a high speed or a heavy load at a low speed. Analysis of the magnitude and duration of the torque required for each winding system will determine the most economic arrangement.

(26) The winding ropes are attached one at each end of the drum barrels and arranged to coil on the drum in opposite directions, so that when the drum rotates one cage will be raised and the other lowered.

(27) In the longwall system a continuous line of working face advances in one direction; the face may be straight or curved.

(28) Work volume to open deposits of minerals is significantly reduced. It is not necessary to construct access routes, power lines, or worker communities, nor is it necessary to take agricultural land out of use and later restore it.

(29) Ventilation causes fresh air to be pulled in through the downcast and passed through the mine.

(30) A niche is a precut at face end, one web deep and a shearer's length long.

(31) Shields, a new entry in the early seventies, are characterized by the addition of a caving shield at the rear end between the base and the canopy.

(32) Compared with shearer, the plow is simpler and easier to operate.

(33) Further sagging causes the blocks to form a semiarch, with the gob end eventually rested on the gob.

(34) If permitted, the props under the face end of the bars may not be set until after the cutting machine has passed, but the difficulty is overcome by a needling the face end of the bars into the top of the coal-seam pud, then set immediately after the cutting machine has passed.

(35) It should be recognized that, in spite of all the efforts to produce a flameless explosive, this has been impossible, and that no explosive is absolutely flameless and none is absolutely safe in inflammable mixtures of firedamp and air, so that the firing of shots in the presence of firedamp is a dangerous practice to be avoided at all times and is forbidden.

(36) The casing which surrounds the impeller directs the flow of the water to the discharge pipe.

(37) It is fully recognized that it is better to rely upon adequate ventilation than upon safety lamps, explosives and flameproof electrical apparatus as safeguards against the danger of explosion which arises from imperfect ventilation.

(38) It is generally conceded that surface mining is more advantageous than underground mining in terms of recovery, grade control, economy, flexibility of operation, safety, and the working environment.

(39) There is no question that the development of mineral resources does impact the environment.

(40) Any safety program should be audited to determine if it is being implemented as designed.

(41) All the factors require that a multiphase and three-dimensional evaluation be made of each potential reserve.

(42) The underground map must be correlated with the surface map in order to show where the underground operations are with respect to the boundary lines and certain surface features of the mineral property.

(43) It is often necessary to know how much current is flowing in a circuit and at what voltage.

(44) Failure can be prevented by ensuring that all wiring is properly insulated.

(45) Another explanation is that the changes brought about by the mine's redistribution of stress trigger latent seismic events, deriving from the strain energy produced by its geological aspects.

(46) Perhaps the most common classification of material is whether the material is metallic or non-metallic.

(47) After the treatment, sewage mixed with the mine water is discharged into the reservoir that is formed by subsidence area.

(48) The following are terms that commonly occur in open pit mine planning.

(49) Accidents that happen in coal mines are basically similar to accidents that happen in other industries.

(50) The lamp is lit by a battery which is fixed on the miner's belt.

(51) As a given mine opening is advanced, spad stations are put in the roof on the centerline which has been established for that opening.

(52) Mining machinery that is too large may be difficult to maneuver and work around in providing auxiliary services.

(53) The length of time the rock safely can be left unsupported—bridge action time—will greatly effect the overall drifting-operation cycle.

(54) It is difficult to determine the resistance against which the fan must operate.

(55) The coal is carried in mine cars which are pulled electrically or by diesel engines.

(56) Explosives are made of substances which, when subjected to sudden intense shock, change into gases which take up many thousands of times the space originally occupied by the explosives.

(57) Laser is the most powerful drilling machine, because there is nothing on earth which can not be drilled by it.

(58) The line brattice is essentially a space divider or temporary partition made of an impervious material that is installed and maintained very carefully and kept as close to the face as possible.

(59) When the width and length of a longwall panel reaches or exceeds a certain value as the face advances, the roof strata in the overburden experience disturbance.

(60) Fuse and caps require cautious handling because the highly sensitive explosive is exposed at the open end of the cap.

(61) The coal is passed into a coke oven and it is heated so that the impurities are removed and the coke is produced.

(62) Mining is stimulated by reduction of the ore stiffness to zero in those areas where mining has occurred, and the resulting stress redistribution to the surrounding pillars may be examined.

(63) High transportation costs, poor communications technology, and a lack of companies that had the capability to invest outside their national boundaries led to an industry dominated by small producers operating on either a local or national scale.

(64) However, it must be noted that as the oxygen content in the air decreases, the lower explosion limit will slowly increase while the upper

explosion limit will drop sharply.

(65) Thompson and Visser (2006) argue that optimal performance of a haul road network can only be achieved through an integrated approach incorporating ① geometric, ② structural, and ③ functional design as well as ④ the adoption of an optimal management and maintenance strategy.

(66) Grindability is the physical property of coal which determines the relative ease of pulverizing or grinding a coal.

(67) There is the surface map which shows the location of topographical features such as coal and other mineral outcrops, streams and rivers, roads and highways, railroads, core drill holes' oil and gas wells, and the boundary lines of the mineral property itself.

(68) Savery's engine, however, could only lift water about 15 m (50 ft), and it remained for Thomas Newcomen in 1712 to develop a pumping engine, consisting of a cylinder and piston connected via an overhead beam to a pump rod.

(69) Backhoe configurations allow for more selective digging and faster cycle times as swing angles can be reduced when loading a truck on a lower level.

(70) Although it is evident that not every mineral can be extracted in this way, in many instances such deposits will respond to economic in-situ leaching.

(71) In an open pit mine, mine development begins from excavation and removal of surface soil, which is called as stripping.

(72) The difference between the weights of the two columns is the pressure in pounds per square foot that produces circulation of the air, and the direction of flow will be toward the column of lesser weight as indicated by the arrows in the figure.

(73) To forget the contribution that mining has made (and continues to make) is to take for granted the significant progress that civilization has made since the last Ice Age. It is also to ignore the fact that the very structure upon which we depend is built on—and with—the products of mining.

(74) There are, however, a number of recent publications which provide considerable information on estimating costs for various sizes and types of mines.

(75) Haul road dust can have a considerable environmental impact, increase maintenance and operations costs, and be a serious safety hazard both in the short term by reducing operator visibility and through long-term exposure, which may cause damage to the respiratory system.

(76) Groote Eylandt is located off the coast of east Arnhem Land in the Gulf of Carpentaria about 640 km from Darwin, Australia.

(77) Most shafts that were constructed in the 1900s were of a rectangular cross-section, because of the shape of the pieces of equipment that were taken down the shaft i. e. cages, skips, and counter weights were all square or rectangular in nature and so it made a lot of sense to sink or mine rectangular shafts.

(78) The net effect of these factors can produce the result shown in Fig. 9. 4, which compares the three portals for a mine producing very large tonnages with a 30-year life where the ground conditions do not require continuous slope support and the depth is between 152 and 304 m.

(79) Coal is a sedimentary rock formed from plants that flourished millions of years ago when tropical swamps covered large areas of the world.

(80) Lush vegetation, such as early club mosses, horsetails and enormous ferns thrived in these swamps.

(81) Layers of mud and sand accumulated over the decomposed plant matter, compressing and hardening the organic material as the sediments deepened.

(82) Over millions of years, deepening sediment layers, known as overburden exerted tremendous heat and pressure on the underlying plant matter, which eventually became coal.

(83) Coal formation began during the Carboniferous Period (known as the first coal age), which spanned 360 million to 290 million years ago.

(84) The greater heat and pressures at these depths produce higher-grade coals such as anthracite and bituminous coals.

(85) After a plant dies and begins to decay on a swamp bottom, hydrogen and oxygen (and smaller amounts of other elements) gradually dissociate from the plant material, increasing its relative carbon content.

(86) Ash includes minerals such as pyrite and marcasite formed from metals that accumulated in the living tissues of the ancient plants.

(87) An inclined shaft has been developed, one belt conveyor (belt width

is 1 000 mm) erected and hoisting changed from the original skip hoisting to belt conveying.

(88) There are two cages in each shafts and when one cage is at the top the other is at the bottom.

(89) The coal is carried in mine cars which are pulled electrically or by diesel engines.

(90) In the past, steam engines were used to pump the water out of the mines.

(91) Gas explosion has a certain concentration scope. Gas can be exploded only when its concentration is in five to sixteen percent and this scope is called gas explosion limit.

(92) Cellulose is the material which reinforces the plant cell walls.

(93) In Wales, there is evidence that the Bronze Age people used coal for funeral pyres, and it is known that the Romans used this fuel.

(94) However, the first practical and consistent use of coal seems to date to England in the Middle Ages.

(95) Following this, other discoveries were made by French and British explorers, but the first recorded actual usage was in Virginia in 1702, where a French settler was granted permission to use coal for his forge.

(96) The first rail transportation was for mining, the first steam locomotive was developed in 1814 by George Stephenson in England for a colliery, and the first electric locomotive was developed in 1883 in Germany for underground use.

(97) The set that moves first is the secondary set, and the set that moves later is the primary set.

(98) In the large-diameter drilling, mud and water are often used as a flushing medium, which is called washing liquid.

(99) There are regular four or six leg chock shields in which all legs are vertical and parallel.

(100) The reason that there is no pump pressure after starting may be that the main valve is not well sealed. The processing method is to repair or replace the main valve.

(二) 汉译英

(1) 人类第一次使用煤的确切时间已被历史所淹没。

(2) 同时，在欧洲和亚洲的浅煤层、浅土层矿井中，长臂开采法还是占据着统治地位。

(3) 两个滚筒相距大约 7～10 米远。

(4) 在小型设备上，转子可以铸成一体。

(5) 在矿井内，无论用于何种排水目的，都可以采用固定式水泵。

(6) 综合推进式与后退式开采系统用于更深层和瓦斯煤矿居多。

(7) 长壁开采技术，根据生产能力和各项技术的经济条件，通常可以分为三类。

(8) 热量高的煤矸石可以采用气化的方法获得煤气。

(9) 如何保护煤矿周边的生态环境，已经引起了人们的高度重视。

(10) 瓦斯对大气臭氧层的破坏是二氧化碳气体对臭氧层破坏的 20 倍。

(11) 井下噪声有强度大、声级高、声源多、衰减慢等特点。

(12) 长壁后退式开采法没有长壁前进式开采法使用普遍。

(13) 最薄的可采煤层大约为 12～14 英寸。

(14) 综合机械化采煤，简称综采。可完成五个工序的机械化。

(15) 按煤矿的地质情况及实际生产状况，机械化可分为三级。

(16) 矿井瓦斯涌出的形式一般分为普通涌出和特殊涌出两种。

(17) 中部凿井法和两端凿井法既省力又省时。

(18) 大多数设备用电力驱动。最常见者为交流电，电压 400～2 200 伏。

(19) 中国煤矿开采技术在过去的 20 年里发展迅速，取得了重要进展。

(20) 只有将煤矿岩石力学与采矿工程结合起来研究才会获得好的结果。

(21) 在美国所有煤田中，特别是阿巴拉契亚南部、科罗拉多州西部和犹他州中部煤田，都发现有断层。

(22) 一块看起来很均匀的岩石试件，却经常被发现它每一厘米的热导率都不一样。

(23) 样品的真密度和比重都能够显示出来。

(24) 结构面产状指结构面在空间的分布状态，由走向、倾角和倾向三要素来表示。

(25) 通过结构网络的计算，获得了微结构面的分形维数。

(26) 变质结构面可以进一步分为残余结构面和再生结构面。

(27) 中国是世界第一产煤大国，煤田分布广，开采条件复杂、多样。

(28) 近年来，煤矿废弃矸石的地面排放所引起的环境污染等问题引起了社会的广泛关注。

(29) 本文对取自中国七个不同地区的煤矸石试样的物理力学特性进行了

分析和评价。

(30) 从长远看,提供廉价、充足的充填材料也是保障充填开采技术持续发展的必备条件。

(31) 鉴于煤炭赋存条件的实际,中国煤炭是以地下开采为主。

(32) 为了实现生产和安全的最大效率,每个子系统都有各自的要求。

(33) 放顶煤采煤法是厚煤层主要采煤方法之一,长期存在的回采率低、着火、煤尘与瓦斯聚集等难题一直未能得以解决。

(34) 大地电磁法已经成为煤炭和天然气资源勘探中的重要手段。

(35) 结果表明,这些技术可以促进瓦斯渗流,提高抽采效率。

(36) 当工作面向前推进时,岩层逐渐离层并破断成块。

(37) 与老顶冒落相关的特征是每两个或四个小的周期性顶板来压有一次非常大的周期来压。

(38) 不同支架工作阻力下煤壁的破坏情况如图 8 所示。

(39) 在这种情况下,已经形成了多种不同形式的煤壁锚杆支护。

(40) 在长壁开采中,地表任一点随着工作面的推进而下沉。

(41) 通过日本的实例可以看出,长壁开采地区地表达到完全下沉的时间取决于埋藏深度。

(42) 与一般采高小于 3.5 米的煤层工作面相比,初次来压与周期来压步距均明显减小,来压强度则明显增大,如何确保工作面设备安全平稳运转已经成为一个亟待研究的新课题。

(43) 选择变质程度较低的 WL 煤样进行不同温度条件下煤样的恒温热解实验,分别称取 40 克质量相同的煤样,选择相同粒径的煤样(0.15～0.18 毫米),利用升温炉体将各煤样分别加热到 160、200、240 摄氏度(温度低时,反应后期气体释放量小,测量误差较大)。

(44) 煤的热解过程中产生 CO 和 CO_2 气体的同时会伴随大量活性位点的产生,为了研究煤在热解过程中的气体释放规律即相应活性位点的产生规律,实验进行了煤样在 200 摄氏度条件下的恒温热解实验。

(45) 地下开采的产量比例占 90%,即使是未来露天开采有所发展,地下开采的比例也不会低于 85% ,而且地下开采基本都是长壁开采。

(46) 随着掩护支架的前移,直接顶岩层在采空区将发生垮落冒顶,并且垮落以后的破碎岩层不能沿着开采方向传递水平力。

(47) 煤是植物残骸在覆盖的地层下,经复杂的生物化学和物理化学作用,转化而成的固体可燃有机沉积岩,其灰分一般小于 40%。

(48) 该矿主要有六个采区,本文以五采区 31509 工作面和 32509 工作面为研究区，如图 2 所示。

(49) SR 反映的是应力随时间变化的快慢程度，相较于应力指标，可以更直观地反映出岩石的应力调整状况。

(50) 研究发现，增加液压支架的工作阻力并不能促使工作面前方应力的重新分布，然而液压支架可以提高工作面围岩控制效果，改善工作环境。

二、篇章翻译练习

(一) 英译汉

(1) Passage One

More progress has been made in the technology of mining during the past two decades than in all previous mining history. In longwall coal mining, the coal cut down by the shearer or the plow is transported by the armored face chain conveyor (AFC), which is laid across the full face width, to the headentry T-junction, where it is transferred to the entry belt conveyor through the mobile stage loader. In addition, modern longwall mining employs self-advancing hydraulic powered supports (which will simply be called powered supports) at the face area. The support not only holds up the roof, pushes the face chain conveyor (AFC), and advances itself, but also provides a safe environment for all associated mining activities. Therefore, its successful selection and application are the prerequisite for successful longwall mining. Furthermore, due to the large number of units required, the capital invested for the powered support usually accounts for more than half of the initial capital for a longwall face. Therefore, both from technical and economic points of view, the powered support is a very important piece of equipment in a longwall face.

(2) Passage Two

Coal at the face is cut by either a shearer loader or a plow. The shearer loaders generally ride on the panline of an armored face conveyor (AFC), which is laid on the floor parallel to the faceline. The shearer loader moves by means of a self-contained motorized winch running along a static chain whose ends are anchored at the ends of the face near or at the drive units of the AFC. In the recent models, the shearer loaders are self-propelled along a special track without the aid of the chain. Thus the AFC serves as the track for the cutting machine to move on and as a guide to hold the machine in place. Single or double drums similar in some respect to those used in continuous miners but larger in diameter are mounted on the face side of the shearer loaders. The cutting position of a drum can be fixed or hydraulically adjusted by the ranging arms. The cutting force is provided by the rotating torque available at the axis of the drum. The widths of cuts or webs made by shearer loaders are about 20 to 36 in. (0. 50 to 1. 91 m) wide.

(3) Passage Three

In underground mining, the walls and the roof or back of openings are not usually self-supporting. Although where ore bodies have a quartzite, strong limestone or other competent roof, some very large openings have been excavated open without the aid of artificial support. Natural bedding planes and joints may constitute structural weakness. The moving of the ground during faulting and folding may result in the vicinity. Below shallow depth, the pressure of the overlying rock may cause masses of rock to move into excavations. When an underground opening tends to assume the form of a dome or arch by a progressive breaking off the rock in the roof, caving may cease when this shape is attained. Frequently if support is placed immediately after the excavation is made, the breaking down of the roof may be largely prevented, and the load taken by the support is not great, being far less than the entire weight of the rock from the opening up to the surface. On the other hand, in some, especially deep mines, heavy pressure is evident also on the sides and even in the bottom of excavations. For these reasons some kinds of support for the surrounding rock is often required in underground excavations which are to be kept open to enable mining operations to continue.

(4) Passage Four

The term longwall advancing is applied to those methods in which the direction of advance of the face is away from the shafts outward the boundary. The roadways are formed and maintained through the goaf by building stone pack walls. The space between the pack walls is filled in with refuse to form solid roadside packs which serve to support and regulate the descent of the subsiding roof and to keep the roadways open. The space from which the coal is removed, except for roadways, may be packed either wholly or partially depending largely on the amount of debris available or to be disposed of. In the former case, the face is said to be packed solid, whilst in the latter case packs known as intermediate packs or waste packs, are built in the space between the roadways. This is known as "partial packing" or "strip packing", since the packing is in strips at right angles to the face or parallel with the roadways. In some cases no packs are built other than those at the gate or roadsides, and the roof is allowed to collapse fully in the waste. Then this system is called "the caving system".

(5) Passage Five

There is no doubt that coal is of vegetable origin because coals not only contain recognizable plant remains, but transitions can be found between obvious accumulations of vegetable matter, e. g. peat, through lignites or brown coals, into true coals and into anthracite. This series, from peat, through lignite, into pure coals and finally anthracite is called the "coal series". The position of a coal in the series is termed its "rank". Thus lignite is a very low-rank coal, while at the other extreme, anthracite is a very high-rank coal. Physical appearance, physical properties and chemical composition of coals change with rank as do their utilization characteristics, e. g. calorific value, coking properties and gas generation potential. Thus knowledge of the rank of a coal can be a guide to its utilization.

(6) Passage Six

Under normal circumstances when plants die, they are exposed to air and are broken down primarily by oxidation and also by various organisms, particularly the fungi and aerobic bacteria. Where plant remains accumulate in swamp or bog environment, however, they become water saturated. Aerobic decay soon depletes the water of oxygen, the aerobic organisms die off and anaerobic bacteria take over. The anaerobic bacteria operate without oxygen but they are equally capable of breaking down organic matter as the aerobic forms. Because of the stagnant nature of swamps and bogs, however, the waste products of the bacteria are not flushed away, but build up in interstitial waters and ultimately render the environment sterile. Bacterial activity is thus curtailed and the partially decomposed plant material remains in a state of arrested decay. In this state the material is peat. If the peat is drained the toxic materials are flushed out, decomposition sets in again and the peat may ultimately be destroyed. If the peat is not drained, however, but is buried under relatively impermeable sediments, its geological preservation becomes possible.

（二）汉译英

（1）段落1

大多数煤层看来是原地起源的，也就是说，生成煤的泥炭在原始植物生长和死亡的地方积聚而成。但有些煤的起源却是转移性的，即煤是植物残骸像木筏一样被运移到沉积的地方后形成的。对于原地起源煤，可观察到成煤植物的粗根、细根从煤体延伸到下部的化石土壤——底土岩中的现象。底土岩这东西很有意思，因具有耐火性，故有其本身价值。黏土质底土岩，称为耐火黏土，主要由高岭土矿物（无序高岭石）组成，具有制造耐火砖的利用价值；砂质底土岩（致密硅岩）几乎是纯净的石英岩，可用于制造氧化硅耐火砖。几乎可以肯定，底土岩开始是以普通的黏土和砂粒形式沉积的，但后来上覆泥炭中的腐殖酸浸滤掉了其中的碱性金属并使其结构发生了重组。

（2）段落2

露天坑开采矿山是由于矿石的开采而在地表形成的大坑或凹陷坑，在矿山整个开采年限内都是暴露于地表的。为了揭露和开采矿体，一般需要挖掘和移走大量的废石。任何经营矿山的主要宗旨都是：以尽可能低的成本开采矿床，力求获取尽可能高的利润。露天坑结构参数的选择、矿石与废石的采掘计划是复杂的工程决策问题，有着重大的经济意义。因此，露天坑开采矿山基本上是以经济为指挥棒进行的开采活动，其间受到地质、采矿工程等方面的某些制约。

（3）段落3

在矿井中，有时也必须使用装备在一个容器里的供氧装置。在矿井里，由于井下采区起火，就可能将该区封闭。它的出入口将用厚墙隔绝，以免空气进入和危险的气体逸出。火可能产生大量危险的瓦斯或烟雾，因此人们如果不戴呼吸器，就不能做这个工作。然而，有时也可能有必要重新打开封闭的火区，而在其封闭期间，危险的瓦斯会积聚很多。同时，空气中的氧也可能由于缓慢地或迅速地燃烧而被耗尽，结果，这里的空气通常是不适宜呼吸的。经过专门训练的人员（即救护人员，或正式救护队员，或辅助救护队员）戴着呼吸器，必须首先进入采区巷道，看看必须采取什么措施以使该区安全。这些人把所需要的氧气，像高空飞行员那样，放在氧气瓶里，背在背上，或者把氧气冷却，直至变成液态，然后用呼吸软管送到嘴里使用。

（4）段落4

甲烷存在于煤层内及煤层的围岩（称为“地层”）中。采矿作业中将甲烷泄入矿井巷道中，与空气混合在一起。空气和甲烷的混合气体达到一定的比例就成为爆炸性的。如果你把家里的煤气炉打开，而不立即用火柴点煤气嘴，煤气在炉内将自由地与空气混合。当混合气中空气和煤气达到一定的比例时，你一点火，炉内就要爆炸。在矿井中，如果空气中的沼气量达到爆炸所需的比例，同样可以发生爆炸，不过爆炸规模要大很多。为了防止爆炸，通风必须将沼气稀释到无害的程度，并将其排出矿井。

附录一:综合练习参考答案

一、句子翻译练习参考答案

(一) 英译汉参考答案

(1) 固体充填开采被证实是一个经济上和环境上都可行的方案。

(2) 文章介绍了在复杂地质条件下的隧洞光面爆破施工技术。

(3) 设计相似材料模拟实验的相似参数如下所示。

(4) 因此,可采用国内广泛应用的概率积分法进行固体充填开采沉陷的预计。

(5) 为了避免因采动而影响坝体的稳定性,提高坝体的有效拦水高度,在开采过程中及回采结束后对水库坝体采取了一些措施。

(6) 以实用矿山压力控制理论为指导,结合相似材料模拟实验,研究了煤层开采上覆岩层运动特征。

(7) 岩石饱和吸水率与试件高度几乎呈线性关系。

(8) 指出了建立煤与瓦斯共采理论体系所面临的难题与挑战,展望了煤与瓦斯共采未来的发展方向。

(9) 根据沉积结构面的成因类型分析了它的变形和破坏特征,并且采用理论分析和实验研究了其力学特征及力学效应。

(10) 如果能把井筒定在合理的中央位置,以便尽可能使运输和通风线路的平均长度相等,这无疑是一大优点,但在实际生产中还有一些其他更重要的因素需要考虑。

(11) 过去凿井中费钱最多的是那些需要穿过疏松、流动的含水砂层和弱岩层开凿的工程,由于穿过这些岩层总要增加很多费用。

(12) 然而,必须使井筒具有足够的尺寸,这不仅是为了提升煤炭,也是为了通风的需要。

(13) 现代化装岩作业,可使用一种新型抓岩机完成,这种抓岩机装有一个悬臂,悬臂中央支在枢轴上,能沿着安装在底部吊盘上的单轨道,围绕井筒周边来回移动。

(14) 抓岩机的径向移动、圆周转动以及升降等全部作业,均由风动马达操纵,操作可以在底部吊盘上进行,也可以在吊盘下方的控制室进行。

(15) 在此过程中,含水地层进行人工冻结,然后像坚硬岩石一样进行爆破和挖掘。

(16) 然而,不论何种代用材料,均不具备木材所拥有的独特优点。当负荷大于其强度因而逐渐发生破损时,木材能以一种可以听得到的爆裂声来预告即将来临的毁坏。

(17) 如今,采矿业经常被描绘成一头野兽,它摧毁沿途的一切,为少数人创造财富,为多数人带来持续不断的痛苦。

(18) 在有些情况下,除顺槽或巷道旁侧外,完全不砌筑矸石带,让采空区顶板全部垮落。

(19) 后退式长壁开采法比之于前进式长壁开采法有某些优点,如采空区巷道无须维护,让采空区留在后方。

(20) 滚筒的这种布置,使采煤机在每一次往返行程中均能进行整层截割,从而保证工作面迅速推进并缩短顶板悬露时间。

(21) 如果采用的不是立即支护的支护形式(IFS),截割后的无支护区范围就比较大,顶板悬露的时间也比较长。

(22) 从认识到煤可以燃烧发热,到煤的使用也许就有几千年的历史。

(23) 通风井的隔音室可以建筑在四根架柱之内,另一个办法是把架柱包围在室内。

(24) 提高提升能力可以依靠使用多层罐笼,亦即有两层以上的底板,每层底板携带一个或一个以上的矿车,车数取决于井筒所能容纳的罐笼尺寸。

(25) 可见,对于一定货量,小载荷可以高速提升,大载荷可以低速提升。对每一提升系统所要求之转矩的大小和延续时间进行分析,可以制定出最经济的工作制度。

(26) 两根提升钢丝绳各系结在滚筒的一侧,并且缠绕方向相反,滚筒旋转时,使一个罐笼上升,另一个下降。

(27) 在长壁开采法中,连续的回采工作面以同一方向向前推进。工作面可以是直线的或者弯曲的。

(28) 露天采矿工作量显著减少,这是因为没有必要兴建通道、电源线或工人社区,也不需要将农业用地闲置后再进行土地恢复。

(29) 通风将新鲜空气由进风口吸入,并使之流经井下。

(30) 缺口是在工作面端头预先开出的,其大小为一个截深和一台采煤机机体的长度。

(31) 掩护支架是 70 年代早期新引入的产品,其特点是在底座与顶梁之间的后面部分增加了一个掩护梁。

(32) 与采煤机相比,刨煤机的结构更简单,并易于操作。

(33) 顶板的进一步下沉,导致断裂岩块形成一个半拱,其采空区一端最终支撑在采空区垮落岩石上。

(34) 条件允许的话，在截煤机通过之前，工作面一端顶梁下面的立柱可暂时不设置，但这要通过在顶梁工作面一端煤层的顶部“掏梁窝”来克服困难，等截煤机通过后，便立即安设支柱。

(35) 应当知道，虽然人们竭力想制造出无焰炸药，但至今并未做到，在瓦斯和空气混合所形成的易燃条件下，任何炸药都有危险性，因此，在有瓦斯的情况下放炮，是任何时候都应避免的危险作业，已明令禁止。

(36) 包围叶轮的壳体将水导至排水管道。

(37) 人们已经充分认识到，通风不良有引发爆炸的危险，把防爆的安全措施诉诸良好的通风远比寄托于安全灯、安全炸药、防爆电气设备等更高明。

(38) 众所周知，从回收率、品位控制、经济性、作业的灵活性、安全性及工作环境等角度讲，露天开采都比地下开采好。

(39) 毋庸置疑，矿业资源的发展的确对环境造成了影响。

(40) 任何安全计划都应该进行审查，以确定它是否像制订的那样执行。

(41) 所有这些因素要求对每个潜在储量作多相、三维的评价。

(42) 地下图必须与地面图相关联，以便显示井下作业地点与矿产边界线和某些地表特征的相对位置。

(43) 有必要知道线路中流过的电流有多大，电压有多高。

(44) 防止这种事故的办法是保证导线完全绝缘。

(45) 另外一个解释是，由于矿井应力重新分布带来的变化可能引发潜在的地震，地震是因地质因素产生的应变能释放而造成的。

(46) 也许，材料最常见的分类是根据它是金属还是非金属来进行的。

(47) 污水处理后，连同矿井水混合排入由塌陷区形成的蓄水池。

(48) 下面是一些露天坑开采常用的术语。

(49) 煤矿所发生的事故与其他工业所发生的事故基本类似。

(50) 矿灯由挂在矿工腰带上的蓄电池点亮。

(51) 随着某一矿井巷道往前掘进，测点布置在为该巷道的掘进而设的顶板中心线上。

(52) 过大的采矿机械不易操作，也难于与所提供的辅助设施配套。

(53) 岩石能不支护而又安全的时间即拱桥作用时间，对于整个平巷掘进工艺循环影响极大。

(54) 难以确定扇风机必须克服的阻力。

(55) 煤是用电力或内燃机车牵引的矿车运输的。

(56) 炸药是由致密物质制成的，当受到突然剧烈震动时就变成气体，这些气体要占据数千倍于炸药本身所占的空间。

(57) 激光器是打孔能力最强的钻机,因为它能给世上的任何物体钻孔。

(58) 纵向风障实际上是一种用不透风材料制成的间隔物即临时隔板,所以它要细心安置、精心维护,并使其尽可能接近工作面。

(59) 当工作面长度和推进长度达到和超过一定值的长壁工作面回采以后,煤层顶板上覆岩层会受到扰动。

(60) 导火线接入火雷管的操作需要特别小心,因为高敏感性的炸药已和火雷管敞露端接触了。

(61) 将煤炭送入炼焦炉中加热,以去除杂质并且产出焦炭。

(62) 在采矿区利用减少矿石刚度到零的手段来模拟开采过程,对矿柱周围引起的应力重分布进行检验。

(63) 运输费用高昂,通信技术落后,缺乏实力雄厚的对外投资公司,这些导致了采矿业被作业规模仅在当地或国内的小生产商所掌控。

(64) 然而,必须指出,随着氧气在空气中的含量减少,爆炸的下限将慢慢提高,而爆炸上限则会大幅度地下降。

(65) 汤普森和维瑟(2006)认为只有结合几何形状设计、结构设计和功能设计,并采用最佳的管理和维护策略,才能实现运输路网的最优性能。

(66) 可磨性是煤的物理特性,它决定着煤的相对易破碎性。

(67) 一类是地面图,它表示地貌特征,诸如煤及其他矿物露头、河流、道路、铁路、钻孔、油气井的位置以及矿藏自身的边界线等。

(68) 然而,萨弗里的发动机只能将水提升 15 米(50 英尺),1712 年托马斯·纽科门仍在开发一种水泵,由一个气缸和柱塞组成,通过高架梁与泵杆相连。

(69) 因为挖掘机给卡车进行下装作业时可以减小回转角度,反铲结构带来更灵活的采掘方式,并缩短作业周期时间。

(70) 虽然很明显,不是每种矿物都能以这种方式提取,但在许多情况下,这种矿床会对经济的原地浸出给出反馈。

(71) 在露天矿,矿井开采从开挖和移走表土开始,这就叫作剥离。

(72) 两个空气柱之间的重量差,就是造成空气环流的压力,以每平方英尺的磅数来表示,其流动方向朝着较轻空气柱一方,如图中箭头所示。

(73) 忘记采矿所作的贡献(而且还在继续),就是把自上个冰河时代以来文明所取得的重大进步视为理所当然,也是忽视这样一个事实,即我们所依赖的建筑物就是建立在采矿产品基础上并且是用矿产品建造的。

(74) 然而,最近许多出版物提供了不少有关不同规模和类型矿井费用预算的资料。

(75) 运输道路上的扬尘会对环境造成很大的影响，增加维护和作业成本，是严重的安全隐患。从短期影响来看，扬尘会降低能见度，影响司机的视线；如果长期暴露于扬尘中，会对工作人员的呼吸系统造成极大的损害。

(76) 格鲁特岛位于澳大利亚东海岸的卡奔塔利亚湾阿纳姆地区，距达尔文市约 640 千米。

(77) 多数在 20 世纪建造的井筒都是矩形断面，因为被带下井的设备，如罐笼、箕斗以及平衡锤都是方形或矩形的。因此，掘进和开采矩形矿井很有意义。

(78) 这些因素带来的结果如图 9.4 所示，它是产量很大、有 30 年服务年限矿山三种井筒的比较结果，这里的地层条件不需要连续斜井支护，埋藏深度在 152 米和 304 米之间。

(79) 煤是由数百万年前茂盛的植物形成的沉积岩，那时地球的大部分被热带沼泽所覆盖。

(80) 这些热带沼泽中生长着茂盛的植物，如原始石松、马尾松和高大的蕨类植物。

(81) 一层层泥沙在腐烂的植物上逐渐积累、逐渐加厚，使其压紧压实。

(82) 数百万年后，越来越厚的沉积层，也就是我们熟知的覆盖层，对深埋其下的植物产生高温高压作用，使之最终形成了煤。

(83) 煤的形成起源于石炭纪(也就是我们熟知的第一煤纪)，大约从三亿六千万年前到二亿九千万年前。

(84) 在那个深度，高温高压形成了如无烟煤和烟煤等高质量煤。

(85) 死亡后的植物在沼泽底部开始腐烂时，氢和氧(和一些极少量的微量元素)逐渐从腐烂的植物上分解开来，使含碳量增加。

(86) 灰分中含有诸如黄铁和白铁的矿物质，它们来源于金属，这些金属在古生物未死的组织中沉积。

(87) 开凿了斜井，安装了带式输送机(带宽 1 000 毫米)，由原来的箕斗提升改为了带式输送机输送。

(88) 每个竖井内有两个罐笼。一个罐笼在井口时，另一个罐笼就在井底。

(89) 煤是用电力或内燃机车牵引矿车运输的。

(90) 过去用蒸汽机将水抽出矿井。

(91) 瓦斯爆炸是有一定浓度范围的，只有在 5%～16%这个浓度范围内，瓦斯才能爆炸，这个范围称为瓦斯爆炸界限。

(92) 纤维素是增强细胞壁强度的物质。

(93) 在威尔士，有证据显示在黄铜时代，煤已经用于火烧式葬礼了。而且大家还知道,当时罗马人已经开始使用这种燃料了。

(94) 然而,最开始实际而持续地使用煤的地方应该是在中古时期的英格兰。

(95) 随后,法国和英国的勘探者也有一些其他的发现。1702 年,在弗吉尼亚人们第一次记录煤炭的使用,那里一名法国移民得到了使用煤炭打铁的许可。

(96) 轨道运输也是因为采煤而发明的,1814 年,乔治·斯蒂芬森在一家英格兰煤矿发明了蒸汽机车。1883 年,德国第一次将电力机车用于地下生产。

(97) 先移动的那套液压支柱是副架,后移动的那套是主架。

(98) 在大直径钻井中,经常使用泥浆和清水作为洗井介质,因而把洗井介质简称为洗井液。

(99) 传统的四个或六个支柱的垛式掩护支架的支柱是竖立并平行的。

(100) 泵启动后无压力故障的产生原因是主阀密封不良。故障处理方法是修复或更换主阀。

(二) 汉译英参考答案

(1) The exact date of man's first use of coal is lost in antiquity.

(2) In the meantime, longwall mining continued to be dominant in Europe and Asia because of thin coal and cover.

(3) The two drums are approximately 7-10 m apart.

(4) In the smaller sizes the rotors may be cast in a single piece.

(5) Stationary pumps are used for all types of mine pumping service.

(6) Integrated advance and retreat system is used mostly in deeper and gaseous coalmines.

(7) Longwall mining techniques, in accordance with the production capacities and the technic-economic conditions, can usually be classified into three types.

(8) High-calorie gangue can be used to obtain gas through gasification.

(9) How to protect ecological environment around mine has aroused great attention of people.

(10) The destruction of mash gas for atmospheric ozone is twenty times as bad as the destruction of carbon dioxide.

(11) Underground noise has the characteristics of intensity, high sound level, diverse sound sources and slow decay and so on.

(12) Longwall retreating system is not so popular as longwall advancing.

(13) The thinnest seam that may be worked is about 12 to 14 inch.

(14) Comprehensive mechanical coal mining is short for comprehensive mining, and it can complete the mechanization of five processes.

(15) According to the geological conditions of the coal mine and the actual production situation, in general, the mining mechanization is divided into three levels.

(16) Normally, there are two types of mine gas emission—normal gas emission and special gas emission.

(17) Middle road drilling method and drilling method at both ends save both effort and time.

(18) The majority of installations are electrically driven. A. C. current is used most often, with the voltage ranging from 400 up to 2 200 volts.

(19) In China, coal mining technology has developed rapidly over the past 20 years and tremendous progress has been made.

(20) Significant results can only be achieved by combining rock mechanics and coal mining engineering.

(21) Faults are found in all U. S. coalfields, especially in southern Appalachia, western Colorado, and central Utah.

(22) In an apparently uniform piece of rock it is often found that the thermal conductivity differs from centimeter to centimeter.

(23) True density and specific gravity of samples can be displayed.

(24) The occurrence of structural plane refers to the spatial distribution of structural planes, which can be expressed by the strike, dip angle and trend.

(25) Through the computation of structural networking, the fractal dimensions of the micro-structural surface are determined.

(26) The metamorphic rock structure planes can further be divided into residual structural planes and recrystallization structure surfaces.

(27) China is the largest coal producing country in the world with widely distributed coal reserves, and a variety of complicated mining conditions.

(28) Recently, issues pertaining to surface disposal of coal-mine wastes, particularly environmental pollution, have received wide attention.

(29) In this paper, the physical and mechanical properties of waste rock from seven different mining areas in China have been analysed and assessed.

(30) In the long run, the essential condition for ensuring sustained development requires cheap and abundant backfilling materials.

(31) In consideration of the prevailing conditions of coal reserves, China's coal resources are mainly mined by underground methods.

(32) Each subsystem has its own requirements in order to achieve maximum efficiency of production and safety.

(33) Top coal-caving is one of the main methods applied in thick coal seam mining. The long-standing problems such as a low recovery rate of top coal, spontaneous combustion of residual coal, accumulation of coal bed methane and coal dust have never been resolved.

(34) The magnetotelluric method has become an important method for exploring coal and natural gas resources now.

(35) The results show that this technique can promote gas seepage flow, and improve the efficiency of drainage.

(36) As the face advances, the bedding planes gradually separate and break in blocks.

(37) The special feature associated with the main roof caving is that for every two to four smaller periodic roof weightings, there is a larger periodic roof weighting.

(38) The failure of coal face with different shield support resistance is shown in Fig. 8.

(39) As a result, many different types of rib bolts have been developed.

(40) In the longwall mining, any surface point is subsiding as the face advances.

(41) The case in Japan shows that the time longwall mining area requires to reach subsidence stablization depends on the depth of cover.

(42) Compared with those obtained from the conventional mining height (<3.5 m) operations, the first weighting and periodical weighting intervals decrease considerably, while the weighting intensity grows significantly. How to ensure that equipment runs safely and smoothly is now an urgent problem for new study.

(43) WL coal with low metamorphic grade was selected for isothermal pyrolysis experiment at different temperatures. Coal samples each of 40 g with the same size (0.15-0.18 mm) were heated up to 160℃, 200℃ and 240℃, respectively, via a programmed temperature oven. If the temperature is low, gas emission will be so small in the later phase of reaction that the

measurement error will be huge.

(44) The formation of CO and CO_2 during coal pyrolysis were accompanied by the generation of abundant active sites. To study gas emission laws, also namely producing laws of corresponding active sites, during coal pyrolysis, isothermal pyrolysis experiments were conducted on coal samples at 200℃.

(45) Underground coal mining produced about 90% of total production. Even if in the future open-pit coal mining becomes fully developed, the percentage of underground coal mining production will not be less than 85% of the total. Longwall mining is the most widely used method for underground coal mining.

(46) The immediate roof will cave and fall in the gob following the advance of shield supports. After caving, it will be broken up and cannot transmit the horizontal force along the direction of mining.

(47) Coal is one type of solid combustible organic sedimentary rock formed by plant remains covered by rock strata subject to complex biochemical and physicochemical process and were converted to solid combustible organic sedimentary rocks, the ash content of which is less than 40% in general.

(48) This study focuses on working faces 31509 and 32509 in the No. 5 mining district, which is one of the six main districts in this mine (Fig. 2).

(49) SR reflects the variation rate of stress with time, which could better reflect the stress adjustment condition of rock than stress itself.

(50) It was found that increasing the loading capacity of shield supports in the face area has very little effect on stress distribution in the longwall face and therefore has very little effect on the face stability. However, it does improve ground control in the face area thereby improving working conditions.

二、篇章翻译练习参考答案

(一) 英译汉参考答案

(1) Passage One

在过去二十年中,采煤技术上所取得的成就比以往整个采煤史上所取得的成就还要大。在长壁工作面采煤时,滚筒采煤机或刨煤机采下的煤,通过贯穿工作面全长铺设的铠装刮板输送机(AFC),运送到运输巷丁字形连接处,然后再由移动式转载机载到运输平巷的带式输送机上。此外,现代长壁采煤法在采面采用了自移式液压支架(简称液压支架)。液压支架不仅支撑顶板,推移工作面刮板输送机,自身向前移动,而且为所有相关的采煤活动提供安全的环境。因此,液压支架的合理选择和使用是长壁采煤法成功的先决条件。此外,由于需求量大,投资在液压支架上的资金通常占长壁工作面整个投资的一半还多。所以,无论从技术角度还是从经济角度看,液压支架在长壁工作面中都是一种很重要的设备。

(2) Passage Two

在工作面上用滚筒采煤机或刨煤机割煤。滚筒采煤机一般在铠装工作面刮板输送机的输送机槽上移动。铠装工作面刮板输送机安装在与采面平行的底板上。滚筒采煤机借助自身装有的电动机并沿着固定牵引链运行的小绞车向前移动,固定牵引链两端锚固在工作面两端,靠近或就在工作面铠装刮板运输机的驱动装置上。在最新型的设备中,滚筒采煤机沿着特制的轨道自动推进,而不必借助于链条的帮助。因此,铠装工作面刮板运输机的作用是作为采煤机向前移动的轨道,且保持采煤机在适当位置工作。安装在滚筒采煤机靠近工作面一侧的单滚筒或双滚筒,在某些方面与连续采煤机上的单滚筒或双滚筒相似,但直径大一些。滚筒切割机的截割位置可以通过液压调节摇臂来控制。切割动力由滚筒采煤机主轴上的转动力矩提供。滚筒采煤机的截深大约是 20～36 英寸(0.50～1.91 米)。

(3) Passage Three

在地下开采中,巷道的两帮和顶板,一般要进行人工支护,尽管有些地方的矿体是石英岩、坚固的石灰岩或其他岩性的稳定顶板,因而有些很大断面巷道的开掘不需要使用人工支护。天然的层面和节理可能会形成构造上的弱点。在断层和褶皱的形成过程中地表的移动会导致临近区域的移动。在地下浅处,覆盖岩层的压力可能会造成大量的岩石进入巷道里。井下巷道,由于顶板岩石不断垮落,而趋向呈现穹形顶,即拱顶,这种形状一旦形成,垮落便会停止。如果巷道开掘一完成便立即进行支护,那么顶板垮落的情况,常常在很大程度上可以防止,而且支架所承受的载荷并不大,比起巷道上方至地表的全部重量要小得多。另一方面,在一些矿井中,尤其是深层的矿井中,巷道的两帮和底板的巨大压力也很明显。因此,在那些需要保持畅通,以便在开采工作得以顺利进行的井下巷道中,常常需要对围岩进行某种类型的支护。

(4) Passage Four

前进式长壁开采法是指采面的开采方向从竖井向边界推进的开采方法。在采空区修建石墙构筑并维护井下巷道。石墙之间的空隙用矸石来填充以构筑坚固的矸石墙,用来支护和控制顶板下降,保持巷道畅通。采煤后,所留下的空间,除巷道外，可全部或部分充填，这在很大程度上取决于可以得到或需要处理的矸石量。在前者情况下，工作面全部填充，而后者叫作中间充填带或矸石充填，即在巷道之间修建充填带。此法称为“部分充填法”，即“带状充填法”，因为充填是呈垂直于工作面或平行于巷道的带状形式。在有些情况下，除平巷或巷道旁侧外，完全不砌筑矸石带，让采空区顶板全部垮落。因此这种开采法被叫作“长臂开采法”。

(5) Passage Five

毫无疑问，煤是由植物形成的，因为煤中不仅含有可辨识的植物残留物，而且可观察到明显植物聚集体的若干过渡类型，例如泥炭、褐煤、纯煤、无烟煤等。这个从泥煤经褐煤、纯煤到无烟煤的系列称为煤系。煤在煤系中所处的位置称为煤级。褐煤在煤系中处于很低的位置，而在与此相反的极端，无烟煤处于很高的位置。煤的物理形态、物理性质和化学组分不同，煤级就不同，可利用性（例如其热值、焦化性及煤气化能力等）也就不同。对煤级的认识，可以作为煤的利用指南。

(6) Passage Six

正常情况下，植物死亡后，将暴露于大气中，在氧化和各种生物（特别是真菌、好氧菌）的作用下被分解。但是，当植物体堆积于沼泽环境之中时，就会使它们变得水饱和。好氧菌的分解作用迅速消耗掉水中的氧，于是好氧菌就死亡了，接着厌氧菌开始起作用了，虽然厌氧菌的作用不需要氧，但同样能够像好氧菌那样分解有机体。可是由于沼泽具有呆滞的性质，细菌的排泄物不能被冲走，而是聚集于间隙水之中，最终使得环境变得无菌化了。于是细菌的活动被抑制，已部分分解的植物体就处于停止分解的状态。处于这种状态的植物体就是泥炭。如果疏干泥炭中的水，冲去有毒物质，分解作用将重新开始，泥炭最终将被破坏掉。可是如果泥炭中的水不被疏干，而且埋藏于隔水层之下，则从地质上讲，泥炭就得到了保存。

（二）汉译英参考答案

(1) 段落 1

Most coal seams appear to have been of in-situ origin: that is to say the coal-forming peat accumulated where the plants lived and died. Some coal appears to be of “drift” origin, and the plant remains were transported, e. g. as log rafts, to the site of deposition. An in-site origin is demonstrated by the presence of roots and rootlets that lead down from the coal into an underlying fossil soil or “seat earth”. Seat earths are of interest in their own right, by virtue of their refractory properties. Clayey seat earths, termed fire clays, consist mainly of kaolin minerals (disordered kaolinite) and are of value for manufacture of refractory bricks. Sandy seat earths, "ganisters", are almost pure quartz rocks and are used for producing silica fire bricks. There can be little doubt that the seat earths were deposited as normal clays and sands, but that they were leached of alkali metals and reconstituted by humic acids from the overlying peats.

(2) 段落 2

An open pit mine is an excavation or cut made at the surface of the ground for the purpose of extracting ore and which is open to the surface for the duration of the mine's life. To expose and mine the ore, it is generally necessary to excavate and relocate large quantities of waste rock. The main objective in any commercial mining operation is the exploitation of the mineral deposit at the lowest possible cost with a view of maximizing profits. The selection of physical design parameters and the scheduling of the ore and waste extraction program are complex engineering decisions of enormous economic significance. The planning of an open pit mine is, therefore, basically an exercise in economics, constrained by certain geologic and mining engineering aspects.

(3) 段落 3

A self-contained supply of oxygen sometimes has to be used in mines, too. A mine may have an underground district sealed off because of a fire which has broken out there. Its entrances and exits will be stopped off by thick walls to prevent air getting in and dangerous gases getting out. The fire may be producing so much dangerous gas or smoke that men cannot do this work without wearing breathing apparatus. However, it may be necessary for the district to be re-opened and, in the time it has been closed, dangerous gases will have accumulated. Also the oxygen in the air will probably have been removed by slow or rapid combustion. In consequence the atmosphere will usually be unbreathable. Specially trained rescuers who are either permanent corps or part-time volunteers wearing breathing apparatus, must first enter the workings to find out what has to be done to make the district safe. The oxygen these men need is carried on their backs in the form of bottled oxygen, as used by high altitude pilots or in the form of oxygen cooled until it becomes a liquid and is supplied to their mouths by breathing tubes.

(4) 段落 4

Methane is trapped in the coal seams themselves and in the layers of rock, called "strata" surrounding the coal. Mining operations release the gas into the workings of the mine, where it mixes with the air. In certain proportions the air and gas mixture is explosive. If you turn on the gas oven at home without putting a match to the burner immediately gas will mix freely with the air inside the oven. If you apply a light when the air and gas in the mixture reach certain proportions, an explosion will occur inside the oven. The same thing can happen in a mine—on a much larger scale—if the amount of firedamp in the air is allowed to reach explosive proportions. To prevent that ventilation must dilute the gas to a harmless level and take it out of the mine.

附录二：矿业英语专业术语中英文对照

advance borehole　超前钻孔
advance mining　前进式开采
advance workings　长壁前进式全面开采法
advanced gallery　超前小平巷
advancing longwall　前进式长壁采煤法
air driven mine car loader　风动矿车装载机
air driven rocker loader　压气式铲斗后卸装载机
angle butt weld　斜口对接焊缝；斜对接焊
anthracite culm　无烟煤粉
anthracology　煤炭学
anthracosis　煤肺病；炭肺；矿工黑肺病
approved shot firing apparatus　安全放炮器；耐爆放炮器
attrition mill　对转圆盘式破碎机；擦碎机
auger drilling　螺旋钻孔
back bolting　顶板杆柱支护
back coming　后退回采
back filling system　分段上向充填开采法
back pulling　回采煤柱
back stoping　上向回采法
back stroke　返回行程；逆行程
backdraught　倒转；反风流；逆通风
backhoe　反铲挖土机
backwall injection　井壁背后灌浆
balance pit　平衡重井筒
balanced housing　平衡装置的机架
bank excavation　阶段采掘
bank method of attack　阶段开采法

bank shaft mouth　坚井口
bank work　阶梯开采
bannock　耐火粘土
bar rigged drifter　架式风钻；架式凿岩机
barings　截煤粉
barrier method　柱式开采法
barrier pillar　边界矿柱
barring　顶板支护
barrow pit　手车运输的露天矿
basal cleavage　基解理；主轴面劈理
bearing　煤层走向；轴承
beater pulverizer　锤式粉碎机
bedded vein　层状矿脉
bench height　露天矿阶段高度；台阶高度；工作台高度
bench hole　梯段的下向垂直炮眼
beneficiated ore　精矿
beneficiating method　选矿法
bituminous coal　烟煤；生煤
black cinder　黑炉渣
blackband　黑矿层
blackdamp　窒息瓦斯；窒息性空气
blanket table　平面洗矿台
blast hole　钻孔；炮眼；爆破孔；鼓风口
blast hole drill　凿岩机
blast inlet　鼓风进口
blast layout　装药布置
blast main　鼓风管；主风管
blasthole　炮眼
blasthole bit　炮眼钻头
blasthole collar　炮眼口
blasthole method　深孔爆破开采法
blasting drift　爆破平巷
bleed of gas　瓦斯喷出
bleeder entry　通风平巷

bleeder hole　放气孔；排泄口
bleeder off hole　排放钻孔
bleeder pipe　排出管；放气管；放水管
blind coal　无烟煤；无焰炭
blind galley　独头巷道
blind lead　无露头矿脉
blind outcrop　盲露头
blind shaft　暗井
bloating　炉渣起泡；炉衬膨胀
block brake　闸块式制动器；瓦块式制动器；块闸
block caving method　分段崩落采矿法；矿块崩落法
block mining　分块开采；分区开采
blockhole　炮眼
blockhole blasting　爆破地面大块岩石
blondin　索道；索道起重机；采掘场架空索道
bloom　初轧方坯；大方坯
bloomer　初轧机；开坯机
blower　吹风机；鼓风机
bootleg　哑炮；拒爆炮眼
bore hole　镗孔；炮眼；钻孔
bore plug　钻孔岩样；钻孔土样
borehole charge　钻孔装药
bottom installation　井底车场设备
bottom layout　井底车场布置
bottom level　井底车场标高
bottom loading belt　底带装载式胶带输送机
bottom priming　底部点火；孔底起爆
bowl classifier　浮槽式分级机；分级槽
brattice way　上部巷道
breast and pillar method　房柱式采煤法
breast stoping　全面采矿法
bright coal　亮煤
briquette　团矿；煤球；型煤
briquetting　压块；块状化；煤砖制造

broaching bit　扩孔钻头
bucker　碎矿机
buddle　斜面洗矿槽；斜面固定淘汰盘
bugdusting　清除煤粉
bulk concentrate　整体精矿
bunker conveyor　矿仓输送机；煤舱运输机
Bunsen burner　本生灯；煤气灯
burden calculation　炉料计算
butt entry　煤巷
butt heading　回采平巷
butterfly valve　节流阀；蝶形阀；双瓣阀
by hand level　辅助平巷
by pit　辅助竖井
cable belt conveyor　缆索皮带运输机；钢绳张紧带式输送机
cable crane　缆索起重机
cable scraper　绳拉式矿耙；塔式挖掘机；缆索挖土机；索式铲运机
car coupling　矿车联接
carbon black　烟黑；炭黑
carbon block　炭精块
carbonaceous coal　半无烟煤
caustobiolith　可燃性生物岩
cavernous vein　孔穴矿脉
caving method　崩落开采法
chain coal cutter　链式截煤机
chain pillar　巷道煤柱
chain scraper conveyor　链板输送机
chamber mining　房式开采
chamfer angle　边缘斜截角；倒角；斜切角
chance cone　强斯型圆锥洗煤机
charging installation　装料设备
chats　矿山废石
check cable　安全绳
check door　阻风门
checker brick　格子砖

chimney draught　烟囱通风
chimney fan　出风筒
chimney flue　烟道
chimney hood　排气罩
chinley coal　高级烟煤
chippy cage　辅助罐笼
chippy shaft　辅助竖井
cinder　炉渣；煤渣；炭渣
cinder bed　炉渣床；煤渣床
cinder coal　极劣焦炭
cinder cooler　出渣口冷却箱
cinder dump　废渣堆
cinder yard　堆渣场
circuit test　爆破网路试验
clean gas　净煤气
clean mining　全采；高回收率开采法
clean ore　洗矿
cleaned coal　精煤
cleaner cell　浮游精选机
cleaner tailings　精选尾矿
cleaning plant　选煤厂；清理装置；清选装置；清选机
cleaning rejects　选煤厂废渣
cleaning unit　选煤设备；净化设备
cleans　精煤
cleans ash　精煤灰分
cleap　交错层理
cleek coal　原煤
cleft　裂缝；隙口
coal block　煤柱；块煤
coal briquette　煤砖；煤块
coal bunker　煤库；煤仓
coal chute　放煤溜槽
coal combine　联合采煤机
coal combustibles　煤的可燃成分

coal culm　煤粉
coal deposits　煤炭矿床
coal depot　贮煤场
coal extraction　采煤
coal flotation　浮游选煤
coal grit　煤质砂岩
coal injection　喷吹煤粉
coal leveling bar　平煤杆
coal mill　磨煤机
coal pick　刨煤镐；采煤风镐
coal planer　刨煤机
coal preparating plant　选煤厂
coal pulverizer　煤炭粉碎机；碎煤机
coal pump　煤水泵
coal recoverey drill　螺旋采煤机
coal seam　煤层
coal shovel　铲煤机；煤铲；煤锹
coal slate　煤质板岩
coal slime　煤泥；煤粘泥
coal tar pitch　煤沥青
coal washer　选煤机；洗煤机
coal washery　洗煤厂
coalcutter loader　联合采煤机
coalification　煤化作用
coaly inclusion　煤包体
coaly shale　煤质板岩
cobber　磁选机；选矿机
coefficient of mine air leakage　矿井漏风系数
coefficient of mining　开采率；采动系数
coefficient of recovery　回采率
coke battery　炼焦炉
coke bed　焦床；底焦
coke blast furnace　焦炭高炉
coke breeze　粉焦；碎焦炭

coke briquette　焦炭块
coke gas　焦炉煤气
combined cutter loader　截装机；采煤联合机
commercial bed　可采煤层
commerical field　可采矿床
commercial seam　可采煤层
common drift　共用平巷
common roadway　共用平巷
companion drift　并行平巷
concentrating machine　选矿机
concentrating table　选矿摇床
concentration plant　选矿厂
conveyor mine　输送机化煤矿
conveyor roller　输送机滚筒
conveyor separator　输送机式分级机
counter entry　平行平巷
counter flush boring　反向冲洗钻进
counter gangway　中间运输平巷
counter head　平行平巷
crib ring　井框
cross cut shears　横向剪切机
cross gangway　斜向平巷
cross vein　交错矿脉
crosscut method　横巷采矿法
crown pillar　阶段间矿柱
crozzling coal　炼焦煤
crude gas　脏煤气；原煤气
cutter loader　联合采煤机；截装机
day eye　倾斜探井
dead coal　非炼焦煤
deep working　深井开采
degassing hole　排放瓦斯钻孔
degree of coalification　煤化程度
dense media preparation　重介选矿

deposit discovery　矿床开拓
developing butt　开拓巷道
developing entry　准备平巷
development face　采准工作面
development opening　准备巷道
diagonal slicing　倾斜分层开采法
dipping working　沿倾斜下向开采
double tracked incline　双轨斜井
drawing shaft　提升井
dressing machine　锻钎机；选矿机
drift conveyer　水平坑道运输机
drift drill　架式凿岩机
drift pillar　平巷矿柱
drift way　水平巷道
drifting machine　架式凿岩机
drill mounting　钻机架
drill round　炮眼组
drill water hose　凿岩机供水软管
driving openings　巷道掘进
dry coal preparation　干法选煤
dry concentration　干选；风选
dry cooling　干式熄焦
dummy drift　独头巷道
dummy pass　空轧孔型
dummy roadway　石垛平巷
dummy shaft　暗井
economic stripping ratio　经济剥采比
entry brushing　巷道挑顶
entry pillar　平巷煤柱
escorial　堆渣场
face run　采煤机在工作面移动的时间
face shield　手持焊接面罩
failed hole　已失效的炮眼；死炮眼；不爆炮眼
fancy coal　上等煤；精选煤

fault coal　劣质煤
fiery coal　瓦斯煤
fiery colliery　瓦斯煤矿
final concentrate　最后精矿
final tailings　最终尾矿
fine coal　粉煤
fine concentrate　细粒精矿
fire brake　防火墙
fireboss　瓦斯检查员；通风员
flashing system　充填采矿法
flat back cut and fill method　水平分层充填开采法
flat back method　上向梯段开采法
floor pillar　阶段间矿柱
flotation pulp　浮选矿浆
foliated coal　层状煤
forced block caving　强制分段崩落开采
fringe drift　岩石平巷
full face driving　全断面掘进
full face round　全断面掘进炮眼组
gallery level　运输平巷水平
gallery sheeting　平巷背板
gas blower　瓦斯泄出
gas brust　瓦斯突出
gas calorimeter　煤气量热器
gas down take　煤气降下管
gas issue　瓦斯突出
gaseous and dusty mine　多尘瓦斯矿
gasproof shelter　瓦斯躲避峒
gate end plate　联络平巷内端的转车盘
gate road　采区顺槽；采空区内运输巷
gateroad bunker　采区运输顺槽煤仓
generator furnace　煤气发生炉
goaf degasification　采空区脱气
goaf stowing　采空区充填

gob flushing　采空区水砂充填
going headway　运输平巷
grate stoker　链算加煤机
gravel face　砂矿工作面
gravity concentration　重力选矿
halvans　杂质多的矿石
hard heading　岩石平巷
hole bottom region　钻孔底
home mining　后退回采
horizon mining　多水平开采法
inclined bottom car　底卸式矿车；漏底车
inclined cut-and-fill　倾斜分层充填法
inclined drift　倾斜巷道
inclined seam　倾斜矿层
inclined shaft　斜井
inclined stone drift　倾斜岩石巷道
inclined throat shears　斜口剪切机；斜刃剪切机
incoalation　煤化
inferior coal　劣质煤；低质煤
inflammability　易燃性；可燃性
inflammable coal dust　爆炸性煤炭尘粉
input well　注入井；注水井
instantaneous strain　瞬间应变
intact coal　未采过的煤层；完整的煤层
intake airway　进风巷道
intake horizon　进风水平
interchangeable converter　可换转炉
intermediate crushing　中碎；中级压碎
intermediate drill　中间钻杆
intermediate entry　中间平巷
jig tailings　跳汰选尾矿
joist mill　钢梁轧机；工字钢轧机
kerosine shale　油页岩；块煤
knock burst　岩石突出；矿山震动

kohlenhobel　刨煤机
laboratory coal crusher　实验室型碎煤机
large coal　大块煤
lateral development　水平开拓
lateral drift　侧部平巷
lateral drilling　侧面钻眼；侧向钻井
lateral opening　走向平巷；边孔
layer mining　分层开采
laying reel　中心出料式卷线机；铺料式卷线机
leading place　超前巷道；超前工作区
level cutting　水平掏槽
level development　水平开采
level drift　水平巷道
leveling bar　平煤杆
lift platform　上升平台；升降机平台
lifting winch　提升绞车
light hammer drill　轻冲机钻
little winds　暗竖井
loader conveyor　装载机的运输机
loader discharge conveyor　装载机的卸载输送机
loader drift　装载平巷
loader shovel　装载铲
loading plough　装载刨煤机
loading ramp　装载坡台；装料斜台
lode chamber　矿脉变厚处
lode mining　脉矿开采
long flame coal　长焰煤
long hole benching　深孔台阶式开采
long hole rods　深孔钻杆
longhole drilling　深孔钻凿
longhole method　深孔爆破开采法
longwall retreating　后退式长壁开采
longwall with caving　崩落式长壁开采
low ash coal　低灰分煤

low grade coal　劣质煤
low grade ore　贫矿；低品位矿石
low seam conveyor　薄煤层输送机
low shaft furnace　矮高炉
low velocity explosive　低速炸药
lump coal yield　块煤产量
maiden field　未采的矿区
main and tail rope haulage　头尾绳运输
main conveyor roadway　主输送机巷道
main drift　主平巷
main winding shaft　主提升井
make of refuse　尾矿产量
mammoth mill　大型选矿厂
manless mining　无人开采
map of mine working　采掘工程平面图
mass breaking　大量采矿
meagre coal　贫煤
mechanical breaking　机械回采
mechanical feeder　机械式给矿机
mechanical flotation machine　机械搅拌式浮选机
mechanical grading　机械筛分
mechanical mining　机械化开采
mechanical mole　平巷联合掘进机
mechanical ventilation　机械通风
medium hard coal　中等硬度煤
medium thickness seam　中厚矿层
metallogeny　矿床成因论
methane accumulation　沼气聚集；瓦斯积聚
methane determination　沼气测定
methane emission　沼气泄出
middle band　中矿带
middle flask　中间砂箱
mill hole mining　漏斗采矿
mill housing　轧机机架

mill layout　轧机布置
mill method　漏斗采矿法
mill table　轧机辊道
mill train　轧机机组
milling hole　矿溜子；溜井
milling method　磨矿法
milling ore　可选级矿石；二级矿石
milltailings　选矿尾矿；工场废渣；尾矿；尾砂
minable thickness　可采厚度
minable width　可采宽度
mine air preheating　预热矿井空气
mine air refrigeration　矿井空气冷却
mine cable　矿用电缆
mine climate　矿井气候
mine concession　矿区
mine conditions　矿山条件
mine conveying　矿山运输
mine explosion　矿井爆炸
mine haulage　矿山运输；矿山拖运
mine lighting　矿山照明
mine management　矿山管理
mine model　矿山模型
mine recovery　矿井恢复
mine rescue equipment　矿山救护设备
mine rescue station　矿山救护站
mine resistance　矿井阻力
mine roadway　矿山巷道
mine safety　矿山安全
mine safety appliance　矿山安全设备
mine section　矿区；采区
mine smalls　粉矿
mine static head　矿井通风静压头
mine technical inspection　矿山技术检查；采矿技术监督
mine velocity head　矿井通风速度压头

mine ventilation　矿井通风
mineability　可开采性
mined out space　采空区
mineral assemblage　矿物共生；矿物组合
mineral bearing froth　含矿泡沫
mineral benefication　选矿
mineral composition　矿物组成；无机成分；矿物质
mineral constituent　矿物组成；矿物成分
mineral dust explosion　矿尘爆炸
mineral microscope　矿物学用显微镜
mineral mining　矿物开采；采矿
mineral output　矿物产量
mineral pollution　矿物污染
mineral prospecting　矿物勘探；矿产普查
mineral separation plant　选矿厂
mineral substance　矿物质
mineral wool　矿物棉；矿渣棉
miner's consumption　矿工矽肺病
miner's helmet　矿工安全帽
miner's lamp　矿工灯；安全灯
miner's level　矿用水准仪
mining engineer　采矿工程师
mining engineering　采矿工程
mining equipment　采矿设备；矿山设备
mining explosive　矿用炸药
mining field　采矿区；采矿场
mining geologist　采矿地质师；矿井地质师
mining geology　矿井地质；采矿地质学
mining industry　矿业；采矿业
mining inspection　采矿技术检查
mining intensity　开采强度
mining lease　采矿用地
mining locomotive　矿用机车；矿山机车
mining losses　开采损失

mining machine　矿山机械；采掘机；采矿机械
mining mechanics　矿山力学
mining operation　采矿工作
mining progress　开采进度
mining rate　开采速度
mining raw materials　采矿原料
mining region　采矿区域；采掘区域
mining regulations　采矿规程；矿山技术操作规程
mining right　采矿权；矿业权
mining science　采矿科学
mobile mine drill　自行式矿用钻机
multiple horizon mining　多水平井采
natural cooling　自然冷却
natural draught　自然通风
natural fuel　天然燃料
natural ventilation　自然通风
open air plant　露天装置
open end method　短柱开采法
open pit method　露天开采法
open stope method　空场采矿法
open stope mining　空场采掘
opencast explosive　露天炸药
opencast mining　露天开采
opencast mining machinery　露天采矿机械
opening up by blind shaft　暗井开拓
opening up by inclined shafts　斜井开拓
opening up by vertical shafts　竖井开拓
openings network　巷道网
ore benefication　选矿
ore blending　矿石混匀
ore breaking plant　碎矿车间
ore briquette　团矿
ore charge　矿石料
ore coal briquette　矿煤团块

ore coke ratio　矿焦比
ore concentration　选矿
ore crusher　矿石破碎机；碎矿机
ore distributor　矿石分配器
ore dressing plant　选矿厂
ore extraction　矿石开采
ore leaching　矿石浸出
ore losses　矿石损失
ore preparation　矿石准备
ore reduction　矿石破碎；矿石还原
ore removal　出矿
ore reserves　矿石埋藏量
ore sizing　矿石筛分
ore smelting　矿石熔炼
ore transshipment plant　矿石转运装置
ore washery　洗矿场
oscillating plough　振动式刨煤机
outcrop mine　露头矿
outcrop mining　露头开采
overhand mining　倒台阶采煤法
overhand stope method　上向梯段开采法
overhand stoping　上向梯段回采
panel entry　盘区平巷
panel method　盘区开采法
peat brick　泥炭砖；泥煤砖
peat cutter　泥炭切割机；泥炭切割刀
peat extraction　泥炭采掘
percussive coalcutter　冲击式截煤机
percussive machine drilling　冲击式钻眼
pick mining machine　冲击式截煤机
pillar and breast method　房柱式井采法；房柱式采煤法
pillar method　柱式采矿法
pillar mining　柱式开采法；矿柱回采
pillar residue　残留煤柱；残留矿柱

pillar robbing　回采矿柱；回采煤柱
pillar system　柱式开采法
piston air drill　活塞式风动凿岩机
piston compressor　活塞(式)压缩机
piston drill　活塞式凿岩机
pit barring　矿井支架
pit bottom　井底；井底车场
pit bottom layout　井底布置
pit outline　露天矿场边界
pit quarry　地下运输露天矿
pitch coal　沥青煤；烟煤
plough position indicator　刨煤机位置指示器
plug drilling　冲击凿岩
pneumatic cell　充气式浮选机
pneumatic concentration　风选法
pneumatic concentrator　风力选矿机；风力选煤机
pneumatic conveyor　风动输送机
pneumatic flotation machine　压气式浮选机；充气式浮选机
powder drift　装药巷道；药室
powder loading density　装药密度
powder magazine　炸药库
powder ore　粉碎矿石
powder type explosive　粉状炸药
powdered coal　粉煤
powdered ore　矿粉；粉状矿
prepared reserves　准备储量；准备煤量
prereduction of iron ore　矿石预还原
primary mining　采区准备
primary rocks　原成岩石；原生岩石
prism plough　校柱式刨煤机
productive mine　生产矿山
productive workings　采区
propeller agitator　螺旋桨搅拌器；叶片式搅拌器
propeller fan　螺旋扇风机；螺旋桨式通风机；螺旋桨式鼓风机

proportion of extraction　回采率
prospect hole　探井;勘探钻孔
prospect shaft　勘探立井
prospecting　勘探；探矿
prospecting bore　勘探钻孔
prospecting drift　探矿巷道
prospecting trench　勘探沟；探槽
prospecting work　勘探工作；探矿工程
proved reserves　证实储量；高级储量；已探明矿藏量
proven territory　勘探证实地区
pulp body froth flotation　矿浆泡沫浮选
pulp density　矿浆浓度
pulp dewatering　矿浆脱水
pulp feed line　泥浆供给管线
pulp thickener　矿浆浓缩机
pulp thickening　矿浆浓缩
pulp transport　输矿浆
push feed drill　推进式凿岩机
quarry bank　露天矿的梯段
quarry blasting　采石爆破
quarry operation　露天开采工作
ram tester　撞击试验机；冲击试验机
rapid plough　快速刨煤机
reamer bit　扩孔钻头
rear conveyor　尾端输送机；尾端运输机
reciprocating trough conveyor　摇动式输送机
rectangular shaft　长方形井筒；矩形竖井
regular stoping method　普通回采法；正规回采法
remelting process　重熔法
removable support　可回收的支架；移动式支架
resue method of mining　选别采矿法
retarding conveyor　阻滞输送机；限速输送机；制动输送带
retreating longwall　后退式长壁开采
retreating mining　后退式开采

revolving dump car　旋转卸料车；翻转卸载矿车
rewash refuse　洗渣
rill cut and fill method　倾斜分层充填采煤法
rill cutting　倾斜分层回采
rise heading　天井巷道
rise working　上向开采法
road lining　巷道支架
road maintenance　巷道维修
roadway construction　巷道掘进
roadway junction　巷道交叉点；平巷交叉道口
roadway support　巷道支架
roche separator 湿式带式分选机；湿式带型磁选机
rock crusher　碎石机；岩石压碎机
rock drilling machine　凿岩机；钻岩机
rock ripper　岩石掘进机；挑顶机
rock roller bit　岩石齿轮钻头
rock shaft　充填料井；下放矸石井筒
rod puller　拔钻杆机；钻杆提取器
roll coal crusher　辊式碎煤机
room and pillar mining　房柱式采煤法；房柱式开采
room mining　房式采煤法
rope winch　钢丝绳运输绞车；钢丝绳绞车
rotary air compressor　旋转压缩机；回转式空气压缩机
rotary bucket excavator　斗轮式挖掘机；转式电铲
rotary crusher　旋转压碎机；回转破碎机
rotary rig　旋转式钻机；旋转钻井机
rotary rock drill　旋转钻岩机；回转凿岩机
rough coal　原煤
rough cobbing　拣选大矿石
rough concentrate　粗精矿
rougher cleaner flowsheet　粗选精选流程图
rougher tailings　粗选尾矿
run of mine bin　原煤仓；原矿仓
run of mine coal　原煤

run of mine feed　原矿给料
run of quarry ore　露天开采原矿
safety glasses　护目镜；防护眼镜
safety lamp　安全灯；矿灯
sand cutter　移动式砂处理机；碎砂机
sand flotation process　尾砂浮选法
sand quarry　采砂场
screened ore　筛选矿石
screened refuse　筛出废料
screening washer　洗矿筛；筛洗器
seam dustiness　煤层含尘量
seam gas content　煤层的瓦斯含量
seam inclination　煤层倾斜；矿层倾斜
seam outcrop　煤层露头
seamless tube rolling mill　无缝管轧机
second mining　二次回采
secondary crusher　二次破碎机；二次轧碎机
secondary drilling　二次爆破钻眼
seismic exploration　地震勘探；地震探查
seismic test　地震波试验
seismograph explosives　地震探矿用炸药
selective concentrate　优先浮选精矿
self-feeder　自动给矿机
self-propelled scraper　自动刮土机
semianthracite　半无烟煤
semibituminous coal　半烟煤；半褐煤；半沥青煤
separating cone　圆锥选矿机
sequence of mining　开采程序
service roadway　辅助巷道
service shaft　辅助竖井；副井；措施井
shaft deepening　竖井延深；井筒延深；矿井延深
shaft drilling　井筒钻凿；钻井
shaft top arrangements　矿井地面设备
sheet deposit　层状矿床；席状矿床；矿层

shielding method　掩护筒法;盾盖式开采法
shovel excavator　单斗挖掘机；铲斗式推土机；铲式挖掘机
shovel loader　铲式装载机
shovel mining　机铲开采
shovelling　铲投;铲装
shrinkage method　留矿法
shrinkage stoping　留矿采矿法；仓储采煤法；留矿回采
shuttle car　穿梭式矿车；穿梭式机动矿车；梭车
side dumper　侧卸式翻斗车；侧卸式矿车
side roadway　侧面巷道
single acting compressor　单动压缩机
single bucket excavator　单斗挖掘机
single stall method　单进路柱式采矿法
single stamp mill　单锤捣矿机；单锤碎矿机
sink-float　重介质选矿
sinking advance　凿井速度
sinking headframe　凿井井架
sinking hoist　凿井提升机
sinking hoisting plant　向下掘进用提升绞车
sinking jumbo　凿井钻台；凿井钻车
sinking machine　凿井机；井筒开凿机
slicing and caving　分层崩落开采
slicing and filling　分层充填开采
slide plough　滑动刨煤机
slightly gassy mine　低级瓦斯矿
slime content　煤泥量
slime control　矿泥控制
slime flotation　矿泥浮选
slime separator　矿泥分选机；矿泥分离器
slurry tank　矿泥沉淀池
slurry treatment plant　矿泥处理车间；煤泥处理车间
snowbird mine　季节性煤矿
solution mining　溶液采矿；溶解采矿(学)；溶浸采矿
specimen ore　矿样

spoil bank 废石堆；排土场；堆积废石的老采区
spoil bank conveyor 废石堆输送机
spoil reclaming 尾矿再选
spontaneous caving 自然垮落；自然崩落
spontaneous gas emission 瓦斯突出
spontaneous mine fire 矿内自燃火灾
square-chamber method 方房开采法；方形矿房采矿法
stage working 阶梯式开采
standard mining construction 矿用标准构造
standard quality coal 标准质量煤
steam coal 动力煤；锅炉煤
stope caving method 崩落采矿法
stoped-out area 采空区
stoping-and-filling 充填回采
stoping layout 采矿设计
stoping machine 采矿机
stoping method 回采方法
storage drift 储矿平巷
stratified deposit 成层矿床；成层沉积
strip mining 露天开采；露天剥离
stripping equipment 剥离设备；露天开采设备
sublevel caving system 分段崩落采矿法
subterraneous quarry 地下采石场
subvertical 立井
supply shaft 送料井
surface miner 露天矿矿工；露天采矿机
surface mining 露天开采；表层采矿
surface mining method 露天采矿法
surface workings 露天开采场
tailing elevator 尾矿提升机；尾矿吊斗
tailings disposal plant 尾矿处理装置
top slicing method 下向分层采矿法
track entry 有轨平巷
trackless pit 无轨露天矿

tub changing　矿车调动
twin-entry mining　双平巷开采
twin shafts　双竖井；双立井
two-heading entry　双平巷
underground air　矿内空气
underground blasting　井下爆破
underground method　井下采矿法
underground mining　地下开采；地下采矿(学)
underlay lode　倾斜矿脉
underlying seam　下部煤层
universal coalcutter　通用截煤机
universal mining machine　联合采煤机
unsized ore　未分级矿石；未过筛矿石
uphill opening 天井巷道
upraise drift　上坡平巷
upward digging　向上挖掘
upward mining　上向采矿
vacant place　采空区
ventilating roadway　通风平巷
ventilation network　矿井通风系统
ventilation raise　通风天井
vertical working　垂直巷道
washed ore　精选矿；洗矿
waste roadway　运送充填料的平巷
wet method of mining　水力开采法
winning assembly　采矿机组；采煤机组
workable deposit　可采矿床
workable reserves　可采储量
working bench　开采台阶
working in benches　阶梯式开采
working of lodes　矿脉开采
working seam　开采煤层；开采矿层

参考文献

[1] BAKER M. In other words：a coursebook on translation[M]. Beijing：Foreign Language Teaching and Research Press，2000.

[2] BELL R T. Translation and translating：theory and practice[M]. Beijing：Foreign Language Teaching and Research Press，2001.

[3] CATFORD J C. A linguistic theory of translation[M]. Oxford：Oxford University Press，1965.

[4] MA L Q，GUO J S，LIU W P，et al. Water conservation when mining multiple，thick，closely-spaced coal seams：a case study of mining under Weishan Lake[J]. Mine Water and the Environment，2019，38：643-657.

[5] MA L Q，ZHANG Y，CAO K W，et al. An experimental study on infrared radiation characteristics of sandstone samples under uniaxial loading[J]. Rock Mechanics and Rock Engineering，2019，52：3493-3500.

[6] NEWMARK P，A textbook of translation[M]. Shanghai：Shanghai Foreign Language Education Press，2001.

[7] NEWMARK P，Approaches to translation[M]. Shanghai：Shanghai Foreign Language Education Press，2001.

[8] NIDA E，TABER C R. The theory and practice of translation[M]. Leiden：E. J. Brill，1969.

[9] SNELL-HORNBY M. Translation studies：an integrated approach[M]. Rev. ed. Amsterdam：John Benjamins，1995.

[10] STEFANKO R. Coal mining technology：theory and practice[M]. Society of Mining Engineers of the American Institute of Mining，Metallurgical and Petroleum Engineers Inc.，1983.

[11] WILSS W. The science of translation：problems and methods[M]. Shanghai：Shanghai Foreign Language Education Press，2001.

[12] 巴尔胡达罗夫. 语言与翻译[M]. 蔡毅，虞杰，段京华，编译. 北京：中国对外翻译出版公司，1985.

[13] 陈登，谭琼琳. 英汉翻译实例评析[M]. 长沙：湖南大学出版社，1997.
[14] 陈福康. 中国译学理论史稿[M]. 上海：上海外语教育出版社，1992.
[15] 陈宏薇. 汉英翻译基础[M]. 上海：上海外语教育出版社，2011.
[16] 陈廷祐. 英文汉译技巧[M]. 北京：外语教学与研究出版社，1980.
[17] 杜承南，文军. 中国当代翻译百论[M]. 重庆：重庆大学出版社，1994.
[18] 范存忠. 漫谈翻译[M]//中国对外翻译出版公司. 翻译理论与翻译技巧论文集. 北京：中国对外翻译出版公司，1985.
[19] 范仲英. 实用翻译教程[M]. 北京：外语教学与研究出版社，1994.
[20] 方梦之. 英语汉译实践与技巧[M]. 天津：天津科技翻译出版公司，1994.
[21] 冯庆华. 实用翻译教程[M]. 上海：上海外语教育出版社，1997.
[22] 冯庆华. 实用翻译教程英汉互译[M]. 上海：上海外语教育出版社，1997.
[23] 冯树鉴. 实用英汉翻译技巧[M]. 上海：同济大学出版社，1995.
[24] 冯树鉴. 英汉翻译疑难四十六讲[M]. 杭州：浙江教育出版社，1988.
[25] 冯伟年. 高校英汉翻译实例评析[M]. 西安：西北大学出版社，1996.
[26] 古今明. 英汉翻译基础[M]. 上海：上海外语教育出版社，1997.
[27] 郭建中. 当代美国翻译理论[M]. 武汉：湖北教育出版社，2000.
[28] 郭著章，李庆生. 英汉互译实用教程[M]. 修订本. 武汉：武汉大学出版社，1996.
[29] 韩其顺，王学铭. 英汉科技翻译教程[M]. 上海：上海外语教学出版社，1990.
[30] 宦秉炼. 实用地质、矿业英汉双向查询、翻译与写作宝典[M]. 北京：冶金工业出版社，2013.
[31] 黄龙. 翻译学[M]. 南京：江苏教育出版社，1988.
[32] 黄志安，张英华. 安全工程专业英语[M]. 北京：机械工业出版社，2018.
[33] 蒋国安. 采矿工程英语[M]. 徐州：中国矿业大学出版社，2011.
[34] 江镇华. 英文专利文献阅读入门[M]. 北京：专利文献出版社，1984.
[35] 金隄. 等效翻译探索[M]. 北京：中国对外翻译出版公司，1998.
[36] 靳梅琳. 英汉翻译概要[M]. 天津：南开大学出版社，1995.
[37] 金岳霖. 知识论[M]. 北京：商务印书馆，1983.
[38] 井升华. 英语实用文大全[M]. 南京：译林出版社，1995.
[39] 卡特福德. 翻译的语言学理论[M]. 穆雷，译. 北京：旅游教育出版社，1991.
[40] 柯平. 英汉与汉英翻译教程[M]. 北京：北京大学出版社，1993.
[41] 李长亭. 实用矿业英语口语[M]. 徐州：中国矿业大学出版社，2014.
[42] 李亚舒，严毓棠，张明，等. 科技翻译论著集萃[M]. 北京：中国科学技术出

版社,1994.
[43] 连淑能.论中西思维方式[J].外语与外语教学,2002(2):40-46,63-64.
[44] 连淑能.英汉对比研究[M].北京:高等教育出版社,1993.
[45] 刘宓庆.文体与翻译[M].北京:中国对外翻译出版公司,1986.
[46] 刘宓庆.现代翻译理论[M].南昌:江西教育出版社,1990.
[47] 刘全福.语境分析与褒贬语义取向[J].中国科技翻译,1999(3):2-5.
[48] 刘永立,尹小军.采矿工程专业英语[M].徐州:中国矿业大学出版社,2012.
[49] 罗新璋.翻译论集[M].北京:商务印书馆,1984.
[50] 吕瑞昌.汉英翻译教程[M].西安:陕西人民出版,1983.
[51] 马祖毅.中国翻译简史:"五四"以前部分[M].2版.北京:中国对外翻译出版公司,1998.
[52] 孟广龄.翻译理论与技巧新编[M].北京:北京师范大学出版社,1990.
[53] 潘荣成,房立政.英汉·汉英矿业工程技术词汇手册[M].上海:上海外语教育出版社,2012.
[54] 彭启良.翻译与比较[M].北京:商务印书馆,1980.
[55] 钱歌川.翻译的技巧[M].北京:商务印书馆,1981.
[56] 冉生涛.实用矿业英语[M].徐州:中国矿业大学出版社,2014.
[57] 沈季良,陈俊芳.矿业英语注释读物:井巷工程[M].北京:煤炭工业出版社,1981.
[58] 孙致礼.翻译:理论与实践探索[M].南京:译林出版社,1999.
[59] 孙致礼.新编英汉翻译教程[M].上海:上海外语教育出版社,2003.
[60] 谭载喜.西方翻译简史[M].北京:商务印书馆,1991.
[61] 泰特勒.论翻译的原则[M].北京:外语教学与研究出版社,2007.
[62] 谭载喜.新编奈达论翻译[M].北京:中国对外翻译出版公司,1999.
[63] 倜西,董乐山,张今.英译汉理论与实例[M].北京:北京出版社,1984.
[64] 王会娟,吕淑文,周建芝.翻译技巧与实践教程[M].徐州:中国矿业大学出版社,2008.
[65] 王克非.论翻译研究之分类[J].中国翻译,1997(1):11-13.
[66] 王泉水.科技英语翻译技巧[M].天津:天津科技出版社,1991.
[67] 王伟,刘莉莎.地下工程专业英语[M].武汉:武汉理工大学出版社,2017.
[68] 韦薇,杨寿康.科技英语文体研究[M].长沙:中南大学出版社,2015.
[69] 魏志成.英汉比较翻译教程[M].北京:清华大学出版社,2004.
[70] 许国璋.许国璋论语言[M].北京:外语教学与研究出版社,1991.

[71] 许建平.英汉互译实践与技巧[M].2版.北京:清华大学出版社,2003.
[72] 许钧.当代法国翻译理论[M].武汉:湖北教育出版社,2001.
[73] 杨莉藜.英汉互译教程[M].开封:河南大学出版社,1993.
[74] 叶子南.高级英汉翻译理论与实践[M].北京:清华大学出版社,2001.
[75] 张干周,郭社森.科技英语翻译[M].杭州:浙江大学出版社,2015.
[76] 张吉雄.煤矿岩层控制英文科技论文撰写范例及词汇[M].徐州:中国矿业大学出版社,2017.
[77] 张克礼.英语歧义结构[M].天津:南开大学出版社,1993.
[78] 张培基,喻云根,李宗杰,等.英汉翻译教程[M].上海:上海外语教育出版社,1980.
[79] 章振邦.新编英语语法教程[M].上海:上海外语教育出版社,1995.
[80] 赵世开.汉英对比语法论集[M].上海:上海外语教育出版社,1999.
[81] 赵振才,王廷秀.科技英语翻译常见错误分析[M].北京:国防工业出版社,1990.
[82] 朱晋科,周本友.矿业英语[M].徐州:中国矿业大学出版社,1989.
[83] 朱哲,沈丛.矿业工程概况[M].徐州:中国矿业大学出版社,2017.
[84] 庄绎传.英汉翻译简明教程[M].北京:外语教学与研究出版社,2002.